Jean Pohl (Ed.)

**Research in
Terrestrial
Impact Structures**

International Monograph Series
on Interdisciplinary
Earth Science Research and Applications

Editor
Andreas Vogel, Berlin

Jean Pohl (Ed.)

Research in
Terrestrial
Impact Structures

Friedr. Vieweg & Sohn   Braunschweig/Wiesbaden

CIP-Kurztitelaufnahme der Deutschen Bibliothek

Research in terrestrial impact structures / Jean
Pohl (ed.). — Braunschweig; Wiesbaden: Vieweg,
1987.
   (Earth evolution sciences)
   ISBN 978-3-663-01891-9   ISBN 978-3-663-01889-6 (eBook)
   DOI 10.1007/978-3-663-01889-6
NE: Pohl, Jean [Hrsg.]

Produced by W. Langelüddecke, Braunschweig

ISBN 978-3-663-01891-9

Contents

Editorial

Jean Pohl

Institut für Allgemeine und Angewandte Geophysik der Universität München,
Theresienstr. 41, D-8000 München 2

Introduction

Interest in impact cratering has been growing in the last years since the discovery by
Alvarez et al. (1980) of enriched siderophile elements of cosmic origin at the paleonto-
logical Cretaceous-Tertiary boundary. Possible biological effects of large impacts on Earth
have been discussed many years ago by Kelly and Dachille (1953), Dachille (1962) and
Gallant (1964), for example. However, the discovery of Alvarez et al. (1980) for the first
time gave some experimental evidence for the suggestion that an impact of a large extraterrestrial body could be the primary cause of the biological catastrophe at the Cretaceous-Tertiary boundary.
Other reasons for renewed interest in cratering in the last few years were the discovery
by Voyager I and Voyager II that the icy satellites of Jupiter and Saturn are covered with
impact craters. And then there is also much evidence that some special meteorites found
on Earth might originate from Mars and from the Moon. Impact ejection could have
brought these meteorites into trajectories crossing the orbit of the Earth.
All these discoveries confirm and enlarge the knowledge of the important role of the collision and impact process in the Solar System, and they show how important it is to
understand the impact process in all details and to investigate all possible consequences.
Whereas many aspects of impact structures, especially the morphology, can be studied by
photogeological and other remote sensing methods on other planets and satellites, it
remains to investigators in terrestrial impact structures to try to unravel the details of the
complex process of crater formation.
At present more than 100 impact structures are known on Earth (e.g. Grieve and Robertson
1979, Grieve 1982). Fig. 1 shows the age distribution of the known terrestrial craters
with diameters larger than 4 km for the last 600 Ma. Errors in the age determination, which
are of the order of tens of Ma in some cases, are not indicated. The corresponding impact
energy was calculated using the relationship of Dence, Grieve and Robertson (1977).
Also shown in Fig. 1 is the number of expected impacts in the last 600 Ma in the indicated
diameter interval calculated with the cratering rate of Dence and Grieve 1969 (see also
Grieve 1984). The expected impacts are illustrated by horizontal bars with a random age
distribution. The figure shows that only a very small sample of the terrestrial craters with
diameters larger than about 30 km is known and it is certainly hazardous to deduce any
periodicities for the cratering rate from this small number of data points. Also any
statistical correlation with other geological events must be considered with extreme

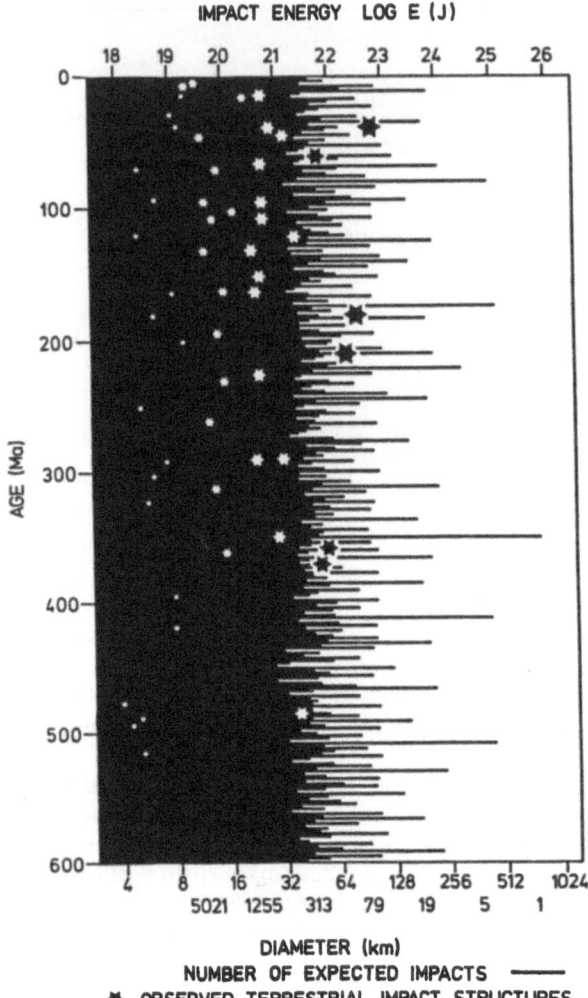

Fig. 1

Age distribution and diameters of known terrestrial impact structures for the last 600 Ma (stars). Number and diameter of expected impacts on Earth in a random distribution (horizontal bars).

caution. Fig. 2 is a representation of the cumulative diameter and the cumulative impact energy of the known terrestrial craters. It illustrates even more drastically that important global catastrophic effects can be expected only from very few of the known impacts. From the above it is evident that it is important to continue the search for terrestrial impact structures in order to enlarge the data base, especially for statistical considerations as a function of time. The cratering rate for the last 600 Ma is already reasonably well constrained by the present data.

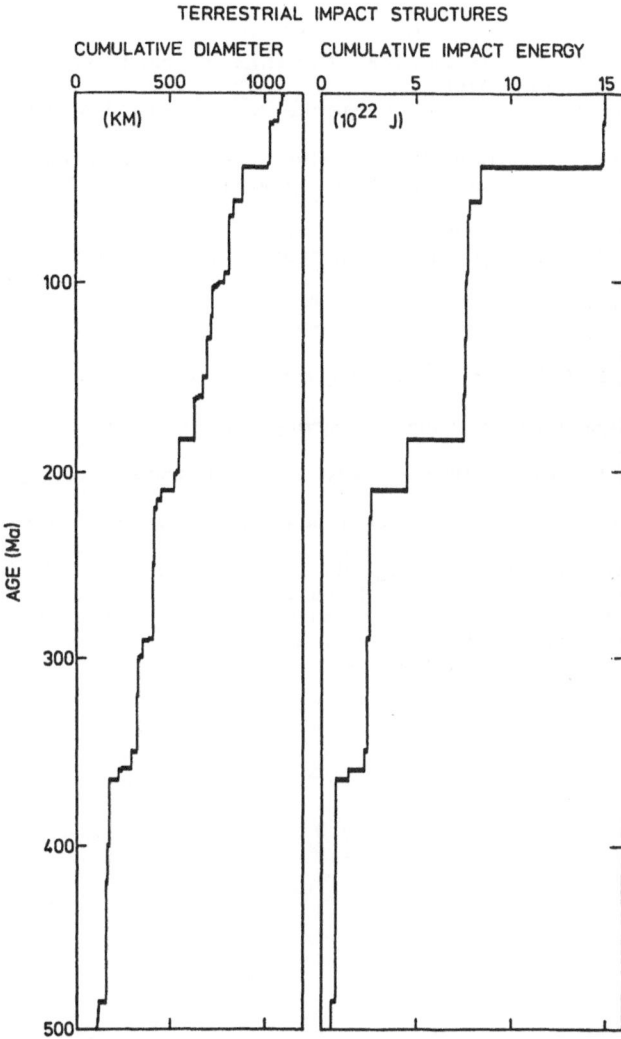

Fig. 2

Cumulative diameters and cumulative impact energies of terrestrial impact structures as a function of age for the last 500 Ma.

The papers collected in this volume address various problems in terrestrial cratering science: Geological description, composition and origin of ejecta, subsurface displacements in the crater bottom and formation of dikes, age determination, and remote sensing methods. In some papers the difficulties and the very detailed interdisciplinary investigations required to try to reconstruct the cratering process clearly show up.

4 J. Pohl

References

Alvarez, W., L. W. Alvarez, F. Asaro and H. V. Michel, 1980: Extraterrestrial cause for the Creta-ceous-Tertiary extinction. Science 208, 1095–1108.

Dachille, F., 1962: Interactions of the Earth with very large meteorites. Bull. South Carol. Acad. Sci. 24.

Dence, M. R., R. A. F. Grieve and P. B. Robertson, 1977, Terrestrial impact structures: Principal characteristics and energy considerations. In: Impact and Explosion Cratering, Roddy, D. J., R. O. Pepin and R. B. Merill, eds., Pergamon Press, New York, 247–275.

Gallant, R., 1964: Bombarded Earth. J. Baker, London, 256 p.

Grieve, R. A. F., 1982: The record of impact on Earth: Implications for a major Cretaceous/Tertiary impact event. In: Geological Implications of Impacts of Large Asteroids and Comets on the Earth, Silver L. T. and P. H. Schultz, eds., Geol. Soc. Am. Spec. Paper 190, 25–38.

Grieve, R. A. F., 1984: The impact cratering rate in recent time. Proc. Lunar Planet. Sci. Conf. 14th, J. Geophys. Res. 89, B403–B408.

Grieve, R. A. F. and M. R. Dence, 1979: The terrestrial cratering record. II. The crater production rate. Icarus 38, 230–242.

Grieve, R. A. F. and P. B. Robertson, 1979: The terrestrial cratering record. I. Current status of observations. Icarus 38, 212–229.

Kelly, A. O. and F. Dachille, 1953: Target Earth. The role of large meteors in Earth science. Target Earth, Carlsbad, Calif., 263 p.

Shoemaker, E. M., 1983: Asteroid and comet bombardement of the Earth. Ann. Rev. Earth Planet. Sci. 11, 461–494.

Shoemaker, E. M., 1984: Large body impacts through geologic time. In: Patterns of Change in Earth Evolution, eds. Holland, H. D. and A. F. Trendall, Springer, Berlin, 15–40.

Whetherill, G. W. and E. M. Shoemaker, 1982: Collision of astronomically observable bodies with the Earth. In: Geological Implications of Impacts of Large Asteroids and Comets on the Earth, Silver L. T. and P. H. Schultz, eds., Geol. Soc. Amer. Spec. Paper 180, 1–13.

Fractures, Pseudotachylite Veins and Breccia Dikes in the Crater Floor of the Rochechouart Impact Structure, SW-France, as Indicators of Crater Forming Processes

Lutz Bischoff / Wolfgang Oskierski

Geologisch-Paläontologisches Institut der Universität, Corrensstraße 24, D-4400 Münster, FRG

Key Words

Impact structures
Rochechouart
Dikes
Pseudotachylites

Received: April 1984

Abstract

Fracture pattern, pseudotachylite veins and breccia dikes were studied in the basement rocks of the Rochechouart structure to clarify their genetic relationship to the impact event. The structural investigations led to the conclusion that the impact-induced fractures in the central crater area are discernible by their greater variety and density compared to the fracture pattern outside the impact structure. But there is clear evidence for the control of newly generated fractures by the pre-existing structural anisotropies.
As expected in crater basements, shear zones moderately dipping to the crater center were detected, indicating that movements of large rock slabs occurred during the modification phase of the developing crater. Related to the shear zones are pseudotachylite veins. For structural, petrographical and geochemical reasons the pseudotachylite melts are evidently generated by the impact event. Besides pseudotachylites, numerous breccia dikes crosscut the target rocks of the crater. It was possible to distinguish different types of breccia dikes, which are obviously related to distinct stages of the cratering process. Finally, it was attempted to correlate the observed phenomena to theoretical cratering models.

1 Introduction

1.1 General Remarks

It is known from theoretical models (e.g. Croft 1977, Maxwell 1977, Melosh 1977) and from crater-forming experiments (Stöffler et al. 1975) which processes take place in the target material at high velocity impact and how they lead to the final morphology of an impact crater.

By extensive field work, carried out in the last decades (e.g. Lambert 1974a, 1977, Robertson & Grieve 1977, Simonds et al. 1978), it has become obvious that some of the terrestrial impact structures are in agreement in their evolution, visible morphology and internal structures with the theories on cratering mechanics while others differ from it. Obviously the differences are due to the anisotropies of geological targets. The crater floor of the impact structure at Rochechouart as well as the surrounding area are composed of highly anisotropic crystalline rocks, i.e. gneisses and micaschists, with numerous intercalations of granite, granodiorite and amphibolite, the whole being cut by various kinds of joints and other fractures (Chevenoy 1958, Raguin 1972).

With our studies we wanted to clarify the question to which extent it might be possible to deduce the impact-induced events from the different brittle deformation structures (joints, faults, breccias), breccia dikes and other related phenomena occurring in the crater floor.

1.2 The Geological Background

The impact structure of Rochechouart, located at the north-west border of the French Massif Central, is a deeply eroded impact crater (diameter about 25 km) of Jurassic age (186±8 Ma, Reimold et al. 1983b). The impact origin was first proposed by Kraut (1967) and later confirmed by Kraut & French (1971), Raguin (1972) and Lambert (1974, 1977). Four different impact formations have to be distinguished: 1) impact melts, 2) suevites, 3) clastic polymict breccias, 4) monomict breccias. Covering an area of about 300 km^2 they form a continuous subhorizontal unit only in a central zone of about 12 km diameter with thicknesses ranging from a few dm up to 60 m.

2 Structural Investigations

2.1 Structural Outline

The principal structural deformation of the Limousin took place during the late Caledonian and Hercynian orogeny. Four distinct deformational steps have been distinguished by Autran & Guillot (1977).

Due to the complex deformation history, the basement rocks are intensely fractured. The main fracture system of submeridional orientation was established in Hercynian times. Deformation along this predominant submeridian direction continued to be active all through the Mesozoic and is related to the formation of the "Bassin d'Aquitaine". Three other major fracture systems striking NW-SE, NE-SW and ENE-WSW can be identified as well (Lambert 1974b), especially on satellite (Landsat) images. Movements parallel to the described directions also involved the subhorizontal layers of impact breccias of the Rochechouart area (Lambert 1974a, 1977; Oskierski 1983). Concentric fractures, well known from other large terrestrial impact structures (e.g. Clearwater, Manicouagan, Siljan), can neither be seen in satellite images nor be proven by field work (Lambert 1974, a,b). There are so far no structural proofs for the existence of a central uplift in the crater floor of the Rochechouart structure, a fact which causes problems with interpreting the cratering mechanics (Lambert 1974b, 1982).

2.2 Fractures in the Crystalline Basement

In all publications about ancient crater floors of impact structures, the authors mention the high degree of fracturing of the target rocks. However, structural investigations have rarely been carried out (e.g. Currie 1972). By structural studies of the basement rocks at the Rochechouart impact structure, we tried to solve the following questions:
— can the influence of the impact event be deduced from the type, orientation and frequencies of joints and other fractures in the crystalline crater floor (comparing statistical investigations)?
— are there any hints as to movements of rock slabs along preexisting or newly generated fault planes caused by the impact (qualitative studies of special regions)?

2.2.1 Statistics of the Regional Fracture Pattern

The interpretation of the joint and fracture pattern is based on the structural analysis of 21 exposures, mostly quarries, in the actual crater floor as well as in its surroundings within a radius of about 26 km, which were probably not influenced by the impact event. The criteria of comparing the fracture pattern within the crater with that of the uninfluenced crystalline basement are: the same types of rocks, similar structural setting before the impact took place, an identical tectonic history after the impact as well as a large number of good exposures. These conditions are not given in the research area. Nevertheless it could be assumed that characteristic impact-related fractures, for instance those of radial or tangential orientation to the impact center, would be reflected by the statistcal diagrams of the fractures.

2.2.1.1 The Crater Region

The pattern of the orientations of all fractures measured at the various locations (contours of frequencies of the fracture plane poles, Schmidt net, lower hemisphere; for contouring intervals see Table I) are shown in Fig. 1 for the crater region, and Fig. 2 for the surroundings of the crater. Lambert (1974 a) subdivided the crater region in an inner and outer zone using as criteria the extension of allochthonous breccias. As can be clearly seen in the orientation diagrams included in Fig. 1, the fracture pattern does not allow the distinction of an inner crater region (diagrams 3, 4 and 5) from an outer crater zone (diagrams 1, 6, 9 and 14), because typical orientation patterns in each zone are missing. Characteristic features, such as identical maxima of joint orientations, radial arrangement of steeply dipping fractures with reference to the impact center, can only be identified on some of the diagrams. It is obvious, however, that in some cases (diagrams 3, 5, 9, 10 and 11) fractures are frequent which dip gently to moderately towards the impact center. So this could indicate a relation between the impact event and the origin of these special kind of fractures. An explanation for the origin of these fractures could be that in consequence of the crater forming processes, blocks of the crater floor or wall were displaced along these fracture planes.
Another hint to an effect of the impact on the fracture system in the crater floor is the great diversity of the fracture orientations, as is clearly demonstrated by the diagrams of the crater floor (e.g. diagram 10). This could mean that the impact triggered the formation of various sets of randomly oriented new fractures in the target rocks in addition to the already existing fractures.

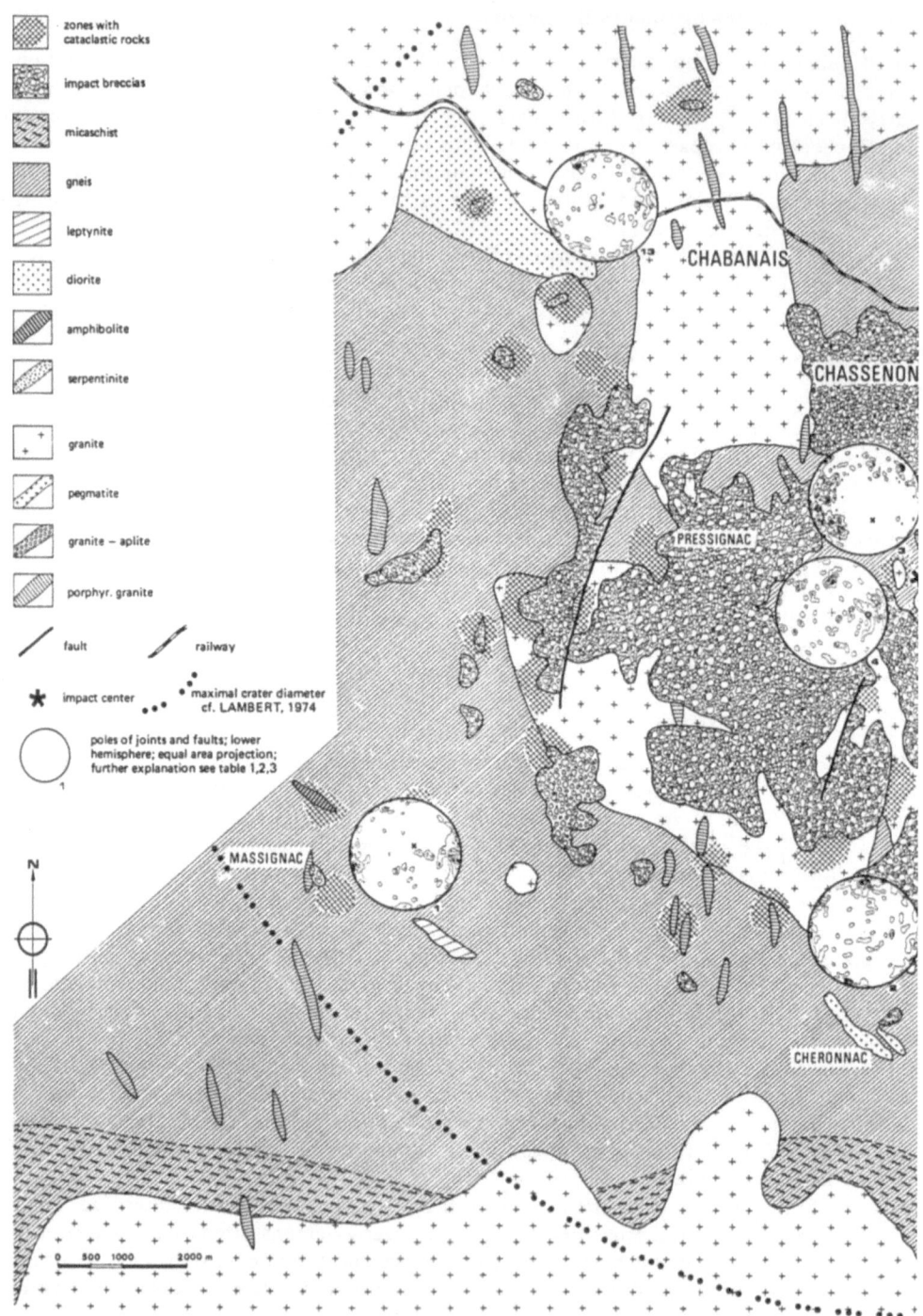

Fig. 1 (pages 8 and 9)
Generalized geological map of the Rochechouart meteorite impact structure (after Lambert 1974) with

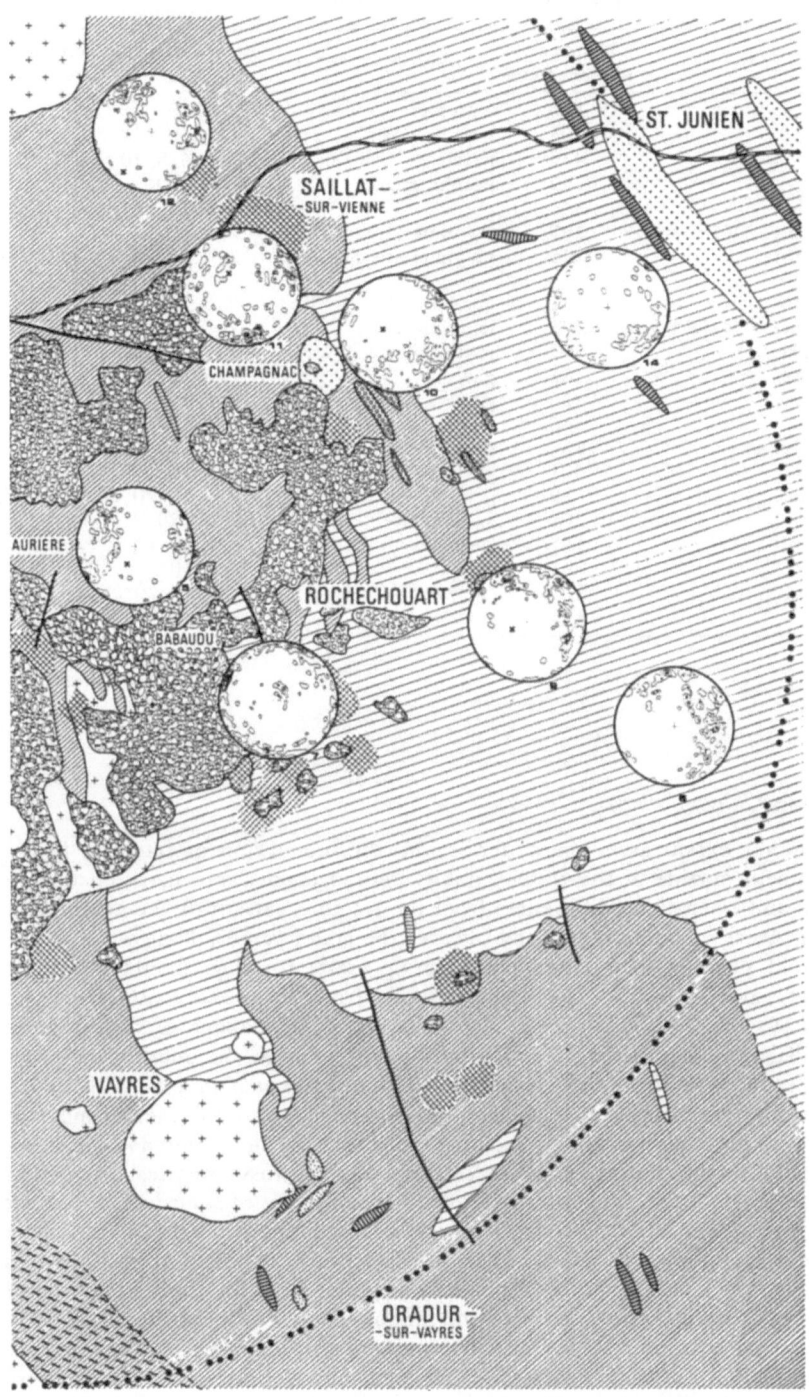

diagrams of the spatial orientation of fractures (joints and faults)

Table I: Statistical data and rock types of the investigated localities

	Locality	Number of measurements	Exposed rock type	Contours in percent per 0.2 area
1	Mazignac	320	leptynite	2, 6, 10
2	Montoume	183	melt breccia	2, 6, 10, 14, 18
3	Lauriere	413	gneiss	2, 4, 6, 8, 10
4	Champonger	149	porphr. granite	2, 10, 20
5	Pierre Folle	343	gneiss	2, 6, 10, 14
6	Rochechouart (D41 East)	297	leptynite	2, 6, 10
7	Rochechouart (castle rock)	245	polymict breccia	2, 6, 10, 18, 22, 26
8	St. Cyr	231	granulite	2, 6, 10
9	Rochechouart (D41 West)	275	leptynite	2, 6, 10, 14
10	Champagnac	395	gneiss, diorite	2, 6, 10, 14
11	Saillat-South	165	gneiss	2, 6, 10, 14, 18
12	Saillat-North	326	gneiss	2, 6, 10
13	Chabanais	340	granite	2, 6, 10, 14
14	Roch.-St. Junien	120	gneiss	2, 10, 20
15	Abjat	401	granite	2, 6, 10
16	St. Mathieu	99	granite	2, 6, 10, 20, 30
17	Champsac	205	amphibolite	2, 10, 20
18	Pagnac	197	gneiss, micaschist	2, 6, 10, 14
19	Péruse	184	gneiss	2, 6, 10
20	St. Germain	272	granite	2, 6, 10, 20, 26
21	Brillac	173	granite	2, 6, 10, 20, 30

2.2.1.2 The Crater Surroundings

To the same minor extent as in the crater region, the orientation of fractures in the crystalline basement outside of the crater exhibits a homogeneous and characteristic pattern. There are also no striking differences from the fracture pattern of the crater floor itself. Fracture planes moderately inclined towards the crater center, along which block movements could possibly have taken place during crater forming, should not occur anymore in this region because of the greater distance from the impact center. However, some diagrams (8, 16, 20, 21) shown in Fig. 2 clearly show maxima of such fracture systems. The diagrams themselves do not supply any answer to the question, whether these fractures only represent locally developed oblique joints or if they are larger glide planes formed during the impact. But field investigations disproved the second possibility; there were no large moderately inclined faults as those encountered in the crater region (see the following sections). Whereas fractures with centrosymmetrical orientation to the crater center only play a subordinate role in the diagrams of some localities, most of the diagrams represent a fairly good correlation to the directions of fotolineations in the immediate neighbourhood (Fig. 3). The fotolineations are taken from Lamberts (1974 b) analysis of a Landsat image. Orientation diagrams from locations situated near or at a prominent fotolineation also show the predominance of this strike direction in the fracture pattern. Diagram 1, for instance shows a maximum of N-S-striking subvertical fractures and a submaximum with NNE strike. The N-S orientated fractures are parallel to several major lineaments (Fig. 3) in the surroundings of the locality. Figure 4 contains data of a quarry, situated very close to a N-S fotolineation. In Fig. 4 breccia

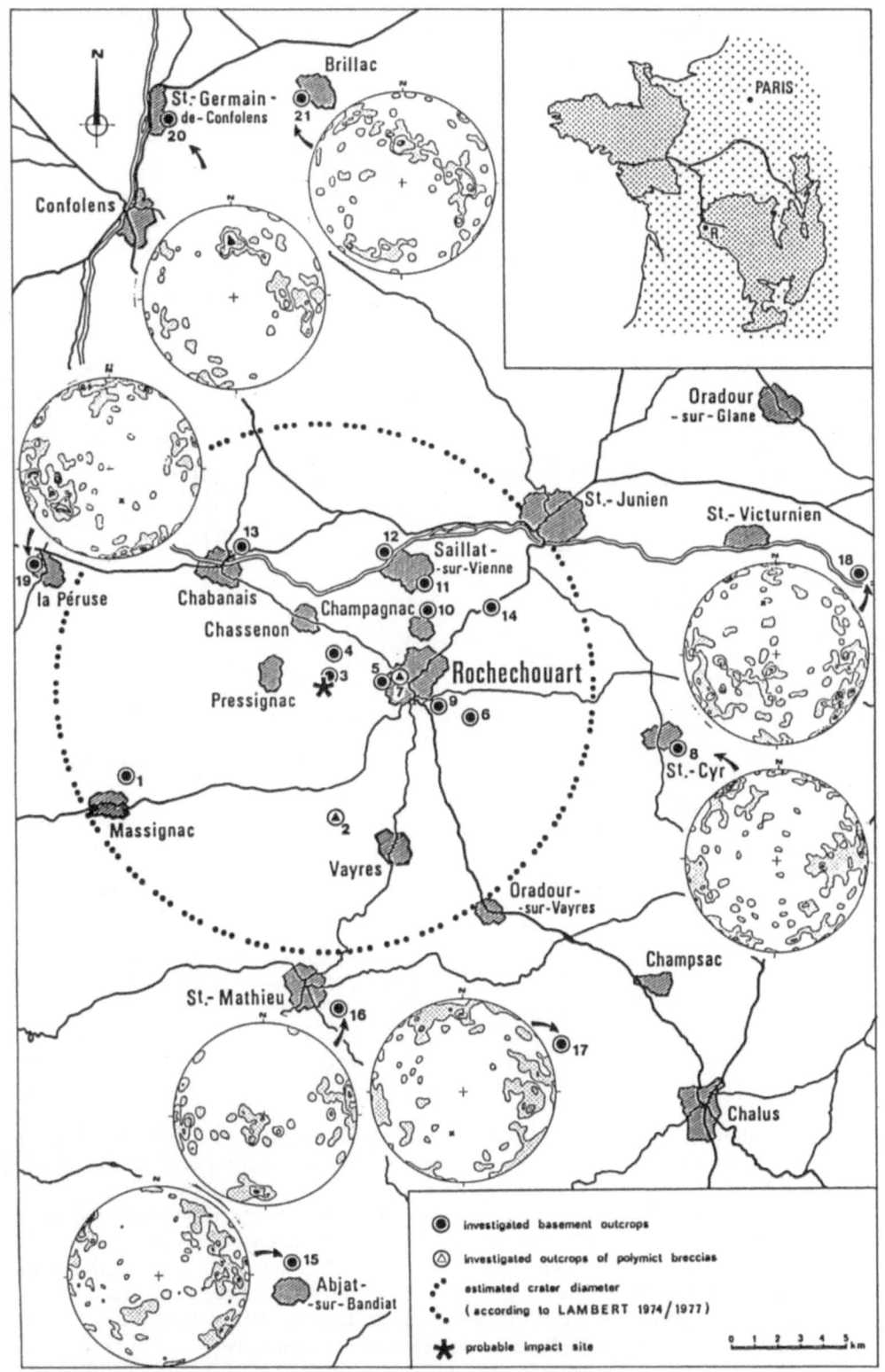

Fig. 2
Fracture orientation (Schmidt net diagrams) in the surroundings of the impact structure.

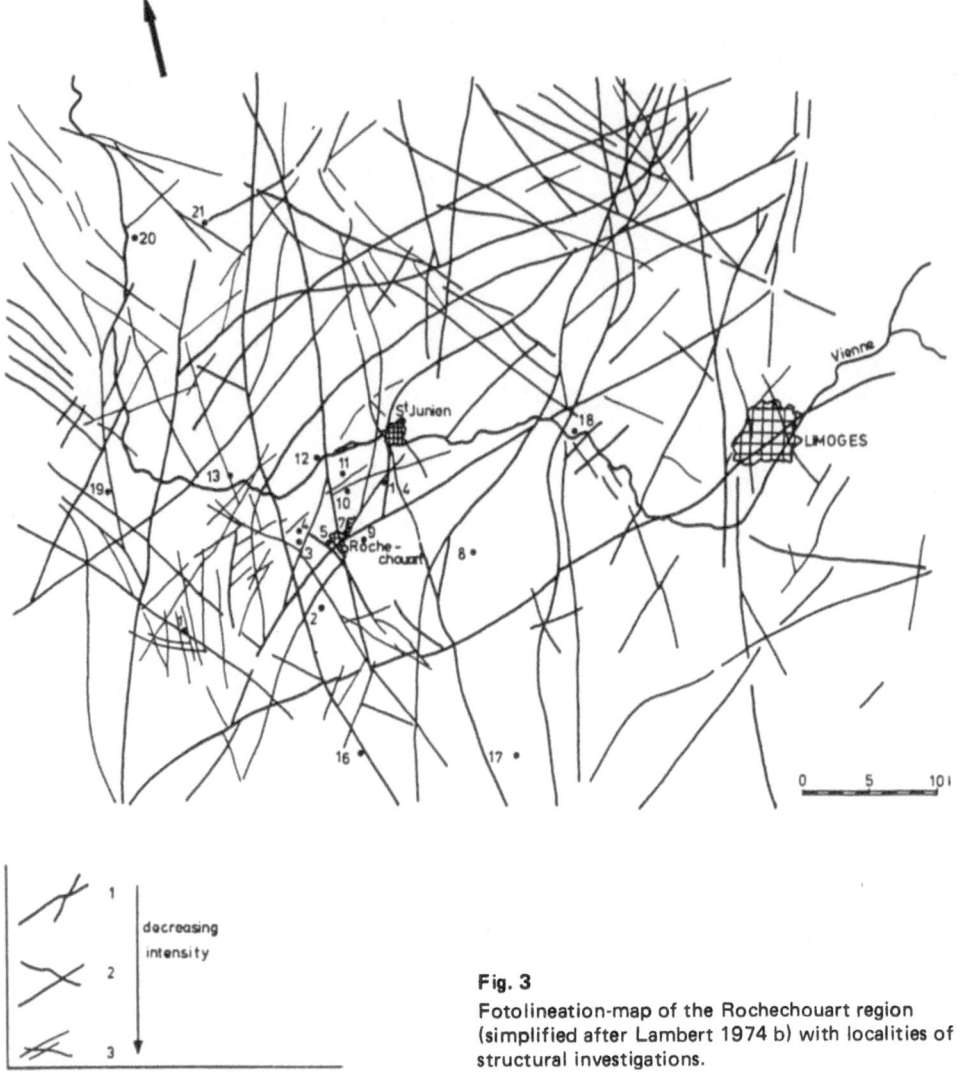

(see Fig. 2)

Fig. 3
Fotolineation-map of the Rochechouart region
(simplified after Lambert 1974 b) with localities of
structural investigations.

dikes also show a preferred N-S direction. Also at locations 16, 17 and 18 (see Fig. 2), an agreement between N-S striking joints and N-S orientated fotolineations can be observed.

Besides the submeridional direction of fotolineaments on satellite images and of joints in the orientation diagrams, many differently orientated fracture systems appear. A correlation between such fracture systems as visible in the diagrams with neighbouring fotolineations of the same orientation cannot clearly be traced. Thus our structural investigation once more supports the results of field research (Lambert 1974 a and others) that the most important fractures in the Rochechouart region are striking N-S.

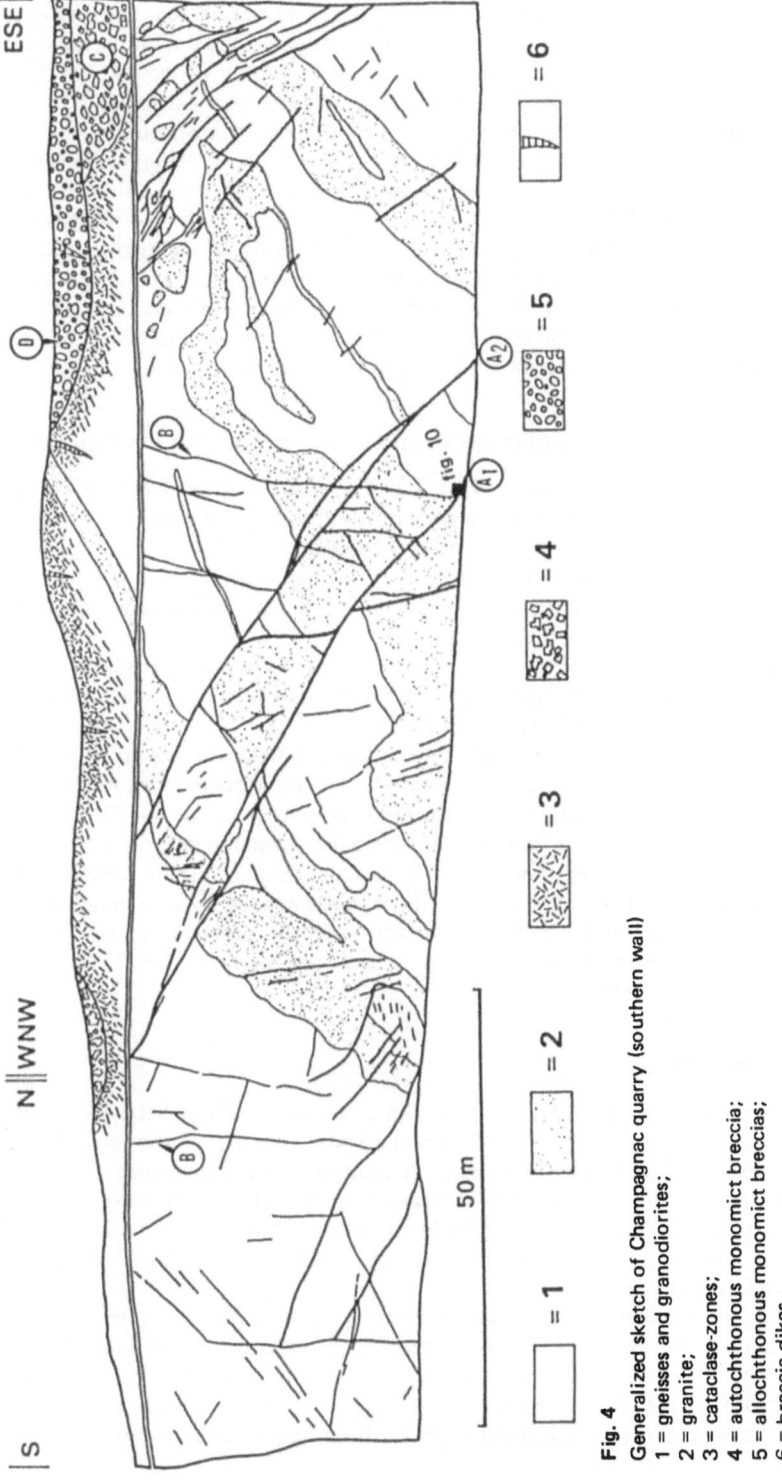

Fig. 4

Generalized sketch of Champagnac quarry (southern wall)

1 = gneisses and granodiorites;
2 = granite;
3 = cataclase-zones;
4 = autochthonous monomict breccia;
5 = allochthonous monomict breccias;
6 = breccia dikes

As a result of the regional fracture analysis, it can be stated that the arrangement of the main joint and fault planes in the entire investigated area within and outside of the impact structure originated mostly from normal tectonic processes and not as a consequence of the impact event. This applies especially to vertical joints. The joint plattern of allochthonous impact breccias (impact melt breccia of Montoume, Fig. 1, diagram 2) and of polymict impact breccia of Rochechouart (Fig. 1, diagram 7), clearly demonstrated by the arrangement of simply conjugated sets of steep joints and a set of subhorizontal fracture planes, supports the idea that they are most probably cooling fissures. The regionally frequent joint directions N-S and E-W are again predominant. This fact can be explained either through a transferring of the fracture pattern below the basement rocks into the breccias, or through the influence of a regional stress field during the cooling of the breccia masses. Some of the gently or moderately inclined fractures in the crater region could probably be the result of the impact process.

2.2.2 Brittle Deformation Structures in the Subcrater Basement at Champagnac

As shown before, the orientation statistics of joints and other fractures do not give any evidence for the influence of the impact on the modification of the fracture pattern. A more detailed structural analysis of several exposures inside the crater provided a better insight into the fracturing of the crater floor during the impact event. This was possible by studying the fracture morphology, movement criteria on fracture planes, relationships of different fractures, and other phenomena.

The greatest variety of impact-related fractures was exposed in a quarry at Champagnac (Fig. 1, location no. 10). It comprises highly fractured rocks, cataclastiç zones, monomict breccias, and several major fault systems.

The following explanations are mostly concerned with the observations made at this location. The main rock type in the quarry are grey gneisses whith a fairly constant schistosity across the whole outcrop, thus indicating that important dislocations probably have not taken place. The gneisses are irregularly intersected by granodiorite bodies, sometime discordant to the schistosity. The youngest intrusions are post-orogenic granites, which are subparallel to the main layering of the gneisses showing pinch and swell structures and ramifying veins. Both the gneisses and the granites are cut by N-S striking, moderately W-dipping quartz veinlets of several cm thickness (Qu-1) with a remarkably constant orientation (Fig. 6).

2.2.2.1 Zones of Cataclasis

Cataclase zones can be found in the whole crater floor. They were described by Raguin (1972) and Lambert (1974 a) and are genetically directly related to the impact event. Within the cataclase zones the density of jointing is extremely high. The spacing of the joints goes down to the cm range. Characteristic features of cataclase zones are generally widened joints but without any visible rotation of the small rock blocks, and also the numerous joint sets of random orientation in addition to the "normal" sets. In general, the open joints are not mineralized.

According to Lambert (1974 a) the cataclase zones, which can reach several decameters of thickness, are restricted to the upper part of the crater floor immediately below the unconformably overlying allochthonous breccia. That also seems to be confirmed by the observations in the Champagnac quarry, where a 1 to 2 m thick cataclase zone is intercalated between an allochthonous breccia sheet and a more or less "normally" jointed gneiss. The intensity of the cataclasis clearly depends on the lithology in the Champagnac quarry; at locations where granites and gneisses crop out below the allochthonous breccia sheet, the gneisses show the typical behavior of cataclastic zones, whereas the granites are brecciated but do not exhibit a dense joint pattern. So it can be stated that generally layered anisotropic rocks more easily develop dense joint systems when affected by an impact shock wave, whereas in contrast more isotropic granites react by brecciating.

2.2.2.2 Monomict Breccias

They are characterized by a fine-grained clastic matrix surrounding more or less unaltered clasts of the parent rock. The clasts differ from the small rock blocks of the cataclastic zones with rounded edges and more irregular forms. The sizes of the clasts and the quantity ratios of clast to matrix vary within a wide range. The degree of brecciation depends on the lithology of the parent rock; while rocks with an internal layering (gneisses and leptynites in the investigated area) do not show a remarkable brecciation, intercalated granite veins may be totally finely crushed. On the other hand, mica-rich gneisses and schists are strongly deformed to breccia-like rocks with aspects of flow structure surrounding more rigid clasts. The degree of dislocation, however, cannot have been very important in some monomict breccias for the metamorphic layering or quartz veins can be traced through the brecciated area (Fig. 5).
Polished and striated surfaces of the clasts testify to the movements which have taken place in the breccias.
The occurrence of the monomict breccias within the intact rocks of the crater floor is highly irregular. At Champagnac these breccias are more abundant and relatively coarser

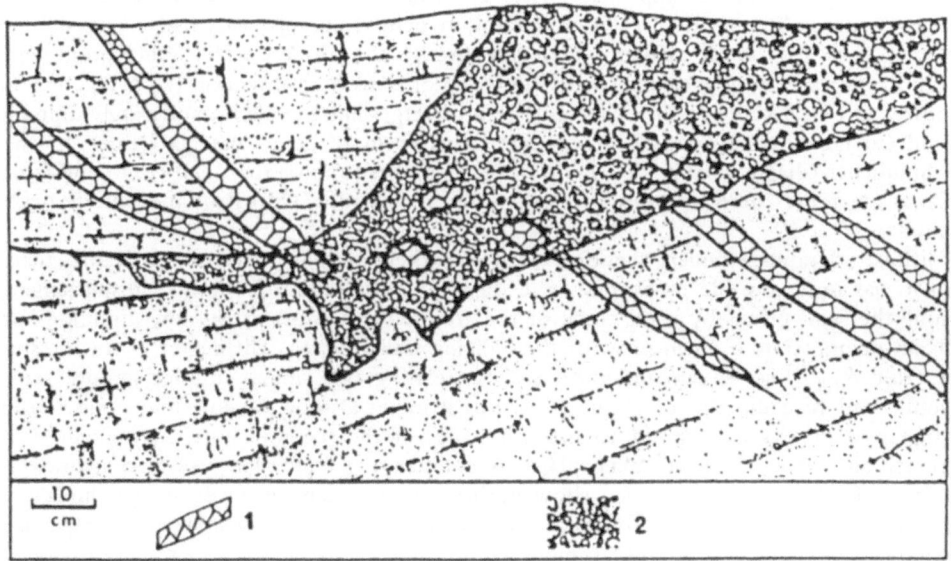

Fig. 5
Autochthonous breccia (2) in gneisses with quartz-I veins (1); Champagnac quarry

in the upper region of the quarry (immediately under the allochthonous breccia cover) but there are no other regularities, neither in their distribution nor in the grain-size variations of the clasts. There seems to be no relationship between clastic zones and monomict breccias, so it cannot be concluded that one kind is a more intensive form of brittle deformation than the other. The cataclastic zones more likely represent an earlier period and another kind of formation (dilatation phase) which is in accordance with Lambert (1974 a, b). The contacts to the intact parent rocks are partly blurred with an interfingering of both rock types, partly abrupt along a joint plane. In a vertical section, repeated intercalations of monomict breccia zones into the normal gneisses can be seen without any visible reason for their occurrence. Anisotropies of the rock fabric, such as joint pattern, faults, orientation of the schistosity, petrographic composition and others, may have played an important role in the formation of the monomict breccias. Today it is almost impossible to find out those prerequisites which facilitated the in-situ brecciation. Only some of the breccia lenses are fixed to important fault planes.

Most of the monomict breccias described so far were autochthonously or subautochthonously developed as a consequence of shock and dilatational waves generated by the impact. The components of other breccias have totally lost the contact to the parent rock and were displaced over greater distances before being redeposited as slabs of monomict brecciated rocks or as rock masses with highly mixed polygenetic clasts. At Champagnac a monomict breccia sheet at least 5 to 6 m thick on top of the whole section (D in Fig. 4) is attributed to this kind of breccia. Its abrupt lower boundary and flat trough-shaped incision into the gneisses with its cataclase zones and its monomict autochthonous breccia lenses, as well as its petrographical composition differing from the underlying rocks, demonstrate its allochthonous character.

2.2.2.3 Joints and Faults

All rock types at the Champagnac quarry are intensively jointed and dissected by fault zones. Among these faults a set of moderately to gently dipping partly listric curved fault planes (A in Fig. 4) are evident which run through the whole quarry. Other prominent fault planes are steeply dipping with submeridional striking direction (B in Fig. 4). As shown on Fig. 4 (upper right), this fault plane diverges in a bundle of moderately dipping auxiliary faults which dissect the granite veins forming broad breccia bands. In the Schmidt net (diagram 10 in Fig. 1), in which the poles to more than 1200 joint and fault planes are represented, these fault systems are clearly reflected by an appropriate maximum.

A lithostratigraphic standard section as well as marker horizons are missing at Champagnac. So it was only possible to estimate the dip separations and directions of displacement along the fault planes by using some of the granite veins. Most of the low dipping fault planes demonstrate a downthrow of the hanging wall (Fig. 8). The degree of the dip slip at fault $A_{1,2}$ (Fig. 4) remains uncertain. Correlating two similar looking granite veins would lead to a dip slip of about 10m. But it might also be possible that none of the hanging wallrocks correspond to the lithologies of the footwall area, so that an even greater dip slip could have taken place. Accompanying fault planes of similar orientations as fault A sometimes even show upthrust of the hanging wall; overall, however, downthrust movements predominate. The steep dipping fault system at Champagnac exhibits only a small or no displacement. Generally they are composed of a set of parallel shear fractures between which feather joints are developed. This fracture system is younger than the listric surfaces (Fig. 10).

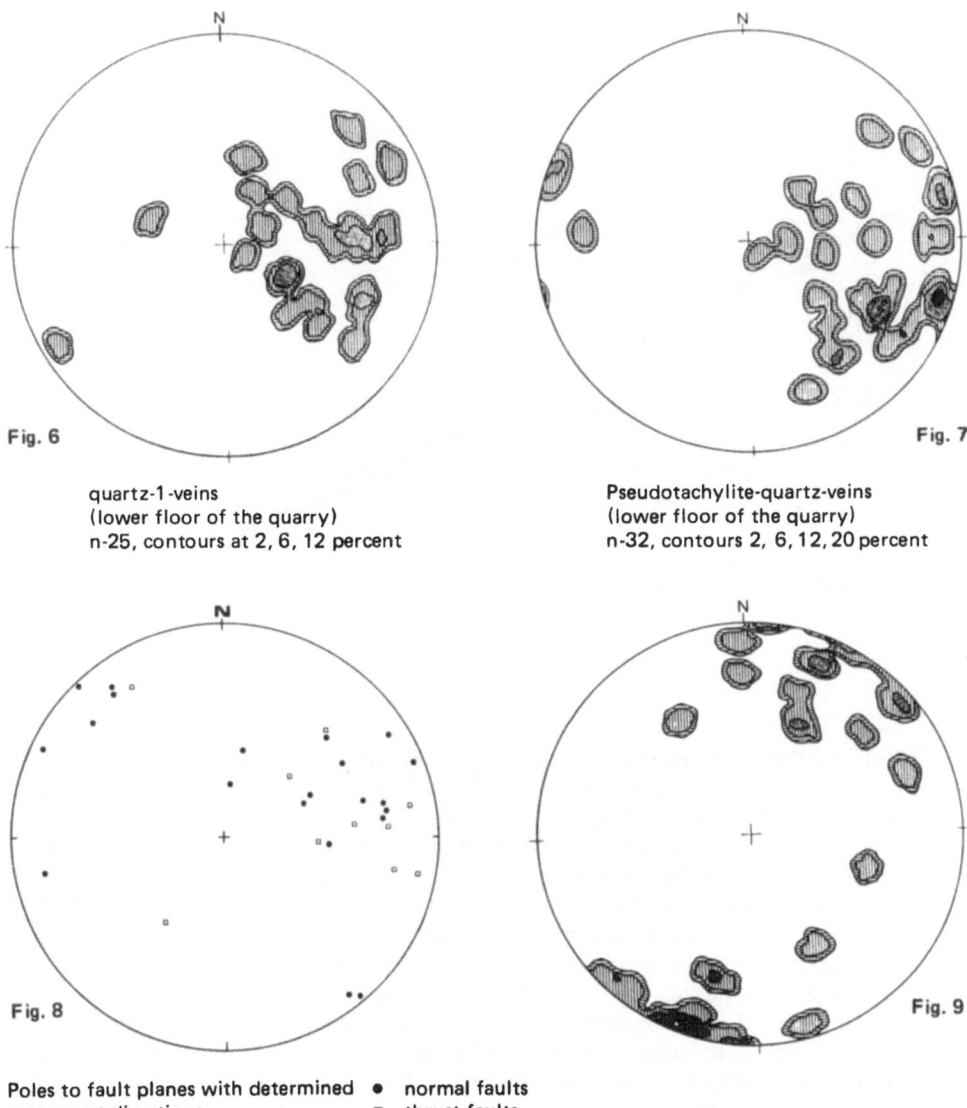

Fig. 6

quartz-1-veins
(lower floor of the quarry)
n-25, contours at 2, 6, 12 percent

Fig. 7

Pseudotachylite-quartz-veins
(lower floor of the quarry)
n-32, contours 2, 6, 12, 20 percent

Fig. 8

Fig. 9

Poles to fault planes with determined   •   normal faults
movement directions   □   thrust faults

Fig. 6—9
Schmidt net diagrams; lower hemisphere; Champagnac quarry.

3 Pseudotachylite-Quartz-Veins

Other particularities in the Champagnac quarry are numerous dark-grey to greenish-grey
aphanitic veinlets of mm to dm thicknesses. Macroscopically they resemble pseudotachylite (PT) veins. Petrographic investigations support their origin from frictional
melting. They are nearly always intimately accompanied by quartz, either as a banding
of quartz- and PT-layers or in a more complicated interfusing of quartz veins with the

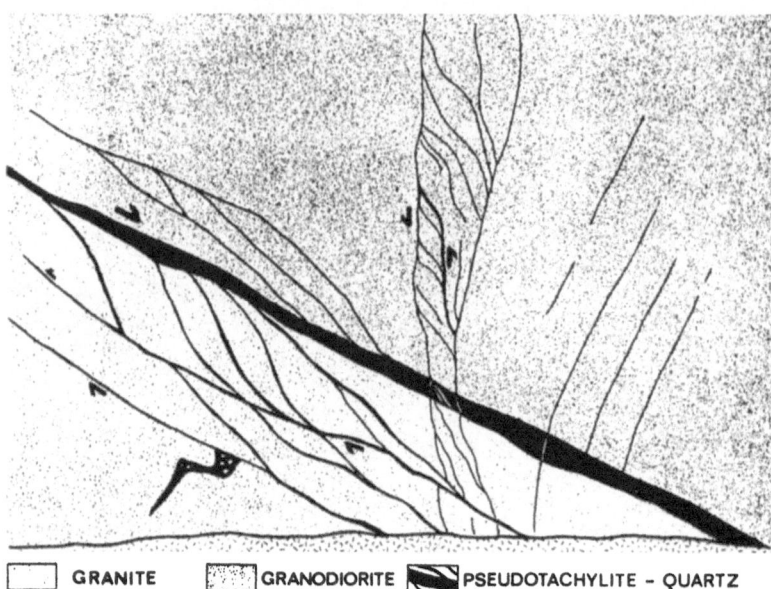

GRANITE   GRANODIORITE   PSEUDOTACHYLITE – QUARTZ

Fig. 10
PT-covered shear fault-planes; detail of Fig. 4 (width 2.5 m)

clearly older PT-veinlets. In thicker veins, small fragments of the wall rock can easily be identified within the PT-quartz-filling, thus pointing out a relationship to brecciation processes. The most evident PT-quartz-veins, which are also the thickest in the quarry, can be traced continuously over more than 60m along the two main faults A_1 and A_2 (Fig. 4). PT-quartz-veins also fill shear faults which are subparallel to those surfaces and the numerous feather joints between them (Figs. 7 and 10). Besides the moderately dipping faults, Fig. 10 shows a system of shear fractures with feather joints without exhibiting an evident degree of slip. These small sigmoidal curved surfaces, too, are covered with PT and quartz. All of the PT-quartz veins mentioned so far are confined to faults. Another kind of PT-quartz veins appears as numerous thin ramifying veinlets forming a network in brecciated granites and granodiorites, less abundant in some gneisses. This net is especially dense immediately below the faults A_1 and A_2.

From the above, it seems to be clear that the PT-vein-filling always took place after the establishment of the fracture systems: PT-quartz-veins never show important fracturing. Even by detailed petrographical investigations, proofs for an in-situ formation of the PT melt could not be found; thus melt probably has been injected into shear fractures, tension gashes, and into voids within brecciated rocks.

A polished hand specimen (Fig. 11) from below the main fault A_2 (Fig. 4) demonstrates the following sequence of microstructural events:

1. Brecciation of the granite
2. Formation of the Qu_1 veinlet (this Qu_1 is not equivalent to the preimpact Qu-1 veins mentioned earlier in the text).
3. Formation of the shear fracture systems s_1 and s_2, displacement of Qu_1. The s_1 surfaces are the most important ones. Other orientations of new fractures are parallel to Qu_1.

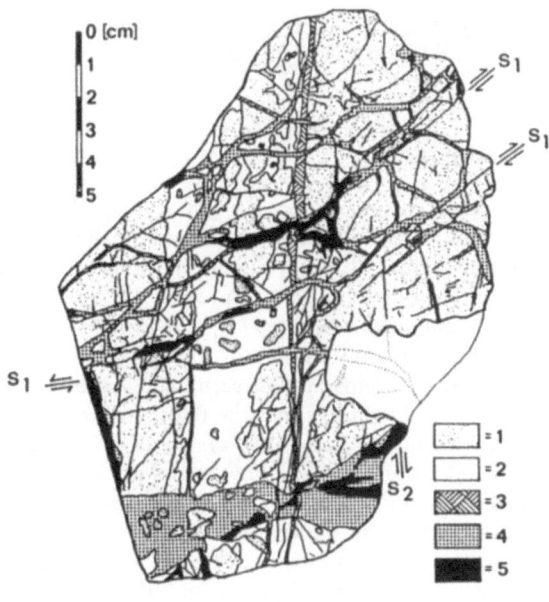

Fig. 11
PT and quartz veins in brecciated granite; hand specimen taken from the shear zone of Fig. 10:
! = weakly brecciated granite; 2 = intensively brecciated granite; 3 = pure quartz vein; 4 = PT-quartz vein; 5 = quartz.

4. The filling of the shear fractures, which were formed at stage 3, with PT and Qu (Qu_2). Qu_2 is also injected parallel to the older Qu_2 veins along hair fissures within the granite and in spotlike enrichments in the intensively crushed zones of the granite.
5. A revival of shear movements along hair fissures parallel to s_2 leads to a small offset of the Qu_2 veins.

The petrography of the PT quartz veins has already been described elsewhere (Reimold et al. 1984 a). On the basis of their composition, two types of veins can be distinguished: Type 1, composed of quartz and sulfides, in which PT-melt in the form of isolated pockets may be included; and Type 2 where PT-melt in the form of long schlieren is again enclosed in a quartz and/or pyrite groundmass. PT-melt shows different forms (glassy, partially chloritized matrix, recrystallized fragmentary matrix, partially recrystallized clasts, altered interstitial glass between quartz crystals). Because of structural reasons, it can be inferred that the PT was formed before the quartz sulfide masses and that a strong dislocation must have taken place after the formation of the PT-melt and before or during the infilling of quartz and sulfides. The quartz of the veins always has a hypidiomorphic to idiomorphic shape, is unstrained and may be considered to be of postdeformational, maybe hydrothermal origin. In all vein fillings the PT-melt components as well as chlorites, which probably were formed out of the PT-melt, are much less frequent than the other vein components.

The PT-melt-bearing veins contain considerable amounts of pneumatolytic-hydrothermal phases. RB-Sr whole rock data of such material align along a rather well defined isochrone of 217±8Ma. It can be excluded for geological reasons that these veins were formed between the end of the regional granitization (dated by Reimold et al., 1983 a to 265±20Ma) and the impact event at 186±8Ma (Reimold et al. 1983 b). It has been shown by Reimold et al. 1984 b that the 217Ma isochrone is of no real geological significance. Evolution of the Rb-Sr isotope system of PT vein fillings was strongly influenced during the pneumatolytic-hydrothermal stage.

Petrographical evidence and the fact that no geological high-energy event is known in the Massif Central lead to the conclusion that the PT veins were formed during the cratering process.

4 Hydrothermal-pneumatolytic Phenomena

Calcite and Ca-, Fe-, Mg-carbonates are abundant in some rocks at Champagnac. They are partly components of the PT-quartz-veins; they partly form monomineralic small veinlets which cut the PT-veins and thus are younger. They also fill vug-like cavities in strong brecciated granites and tension gashes. These tension-gashes show a preferred N-S orientation (Fig. 11) and are perpendicular to the main shear fractures (A_1 and A_2 in Fig. 4). These carbonates are also accompanied by sulfides, mainly pyrite.

Part of the PT-quartz-veins, part of the pure quartz veins and all carbonate-sulfide veins show up to cm thick rims of seritized wall-rocks. In those zones the feldspars are hydrothermally or pneumatolytically disintegrated (Reimold et al. 1984 a). Some of the vein-filling carbonates may come from this disintegration process. Whether an additional infiltration by descending solutions took place remains uncertain. Effects of post-tectonic hydrothermal activity are not only limited to the veins in the crushed granite areas:

— The polymict breccia dikes at Champagnac, compared to other outcrops, are characterized by abnormally high contents of SiO_2 and, in consequence of this, by greater hardness.

— The upper part of the autochthonous gneisses in the Champagnac quarry and its cataclase zones immediately below the overlying allochthonous breccia sheet (D in Fig. 4) are silicified up to 1 m and impregnated by Fe-hydroxides. The breccias above have not been influenced.

— The same monomict breccias (C in Fig. 4) are intersected by a fine network of quartz and fluorite. This clearly demonstrates that the hydrothermal or pneumatolytic events took place after the formation of the monomict breccias.

Reimold et al. 1984 a deduced a temperature below 700 °C from the fact that old muscovite was not altered by the hydrothermal-pneumatolytic processes.

5 Breccia Dikes

Approximately 90 breccia dikes were mapped in the Rochechouart center basement in an area of 80 km^2, which represents nearly a quarter of the whole structure. Modal analyses, textural and structural features of hand specimens, and thin sections lead to the classification of three different types of breccia dikes (Oskierski 1983), which differ in some detail (e.g. recognition of a melt breccia type) from that of Lambert (1981).

Characteristic for Type 1 is the occurrence of mineral and/or lithic fragments surrounded by an impact-melt matrix that has been nearly completely altered to clay minerals. This type of breccia dike only occurs in the center of the structure. Due to characteristic differences in their clast-matrix relationship, two sub-varieties can be distinguished: 1-A generally does not contain lithic clasts. The subrounded mineral clasts have sizes < 0.8 mm, often showing effects of digestion or resorption by the surrounding matrix. Flow structure parallel to the dike walls is quite often visible. Sometimes shock effects up to stage I (according to Stöffler et al. 1971) can be observed. Type 1-A dikes display sharp but irregular contacts with the host rocks. The field appearance of 1-B is very similar to that of 1-A, but dikes of this type are up to 30 cm in width. Contrary to 1-A, rock clasts with maximum sizes up to a few cm are quite common, showing the whole range of shock and thermal metamorphism up to stage III. Flow structures and vesicles are abundant in Type 1-B dikes.

Type 2 dikes are filled by clastic polymict breccias and can be subclassified by clast content and clast size into 2-A and 2-B (matrix of clastic breccias is defined by grain size of constituents $<10\mu m$). A low clastic matrix ratio (<1.5), subangular clasts usually <1 cm in size, and parallel orientation of elongated clasts are typical for the 2-A variety. 2-B displays a significantly higher clast to matrix ratio (>2); clast sizes are up to several cm (sometimes up to several dm). No preffered particle orientation is visible. In both varieties, clasts of Type 1 breccia have been detected frequently (Fig. 12). Type 2-A dikes vary strongly in width, from a few cm up to a maximum of 25 cm. They display a complex geometry and sharp wall contacts (Fig. 13). Type 2-A dikes do not only occur in undisturbed target rocks but also in parautochthonous monomict breccias. Type 2-B has a relatively simple geometry and reaches a maximum thickness of several m. Slickensided contact surfaces can be observed along both Type 2-A and 2-B dikes at several localities and are an indication of tectonic movements. However, displacement of wall rocks was only observed along 2-B dikes.

Type 3 (3-A, 3-B) dikes are characterized by clastic monomict breccia fillings. 3-A contains subangular clasts with sizes <1 cm. The clast to matrix ratio is >2.5 and clasts are frequently aligned parallel to the walls. Sometimes Type 1-A breccia clasts can be observed. Clast sizes of 3-B reach up to several cm. The angular clasts display no preferred orientation and the clast to matrix ratio is much higher (>3.5). Type 3-A is very similar in outcrop appearance to Type 2-A, whereas 3-B (from several cm up to several dm in thickness) does not show sharp contacts with the country rock.

Besides Types 1—3, complex dikes have been found in which the central part of a dike is composed of Type 2-A, and marginal regions of Type 3-A breccia. Between 2-A and

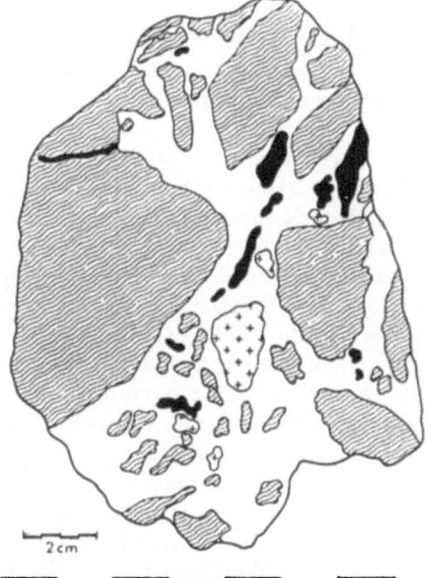

2 cm

Fig. 12

Hand specimen of polymict breccia dike Type 2-B, (locality 3; Laurière-quarry)
1 = gneiss;
2 = granite;
3 = Type 1-A melt breccia;
4 = clastic matrix.

= 1   = 2   = 3   = 4

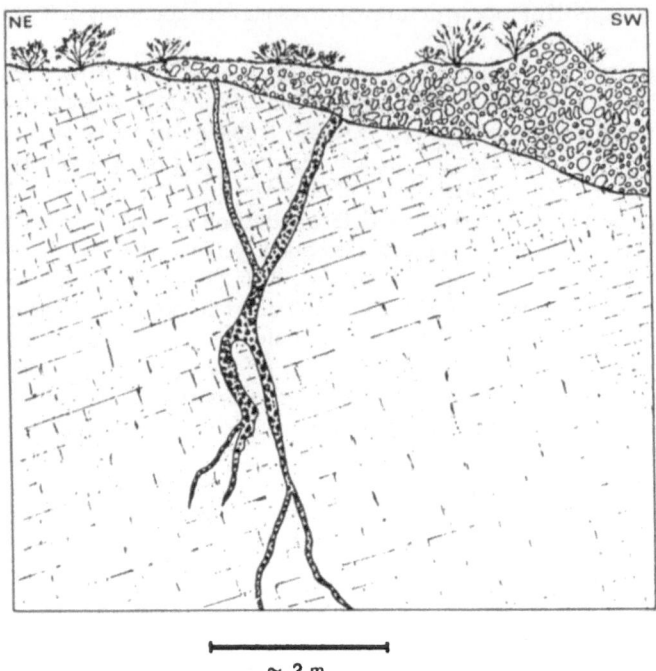

Fig. 13
Type 2-A polymict breccia dike in cataclastic gneisses, superposed by parautochthonous monomict breccia, Champagnac quarry.

3-A, as well as between 3-A and host rock, sharp contacts are visible. Shock metamorphic effects up to Stage I can be found in Type 2 as well as in Type 3 breccia dikes. However, a simple relationship of the shock degree of clasts to the dike positions relative to the point of impact cannot be established.

The chemical compositions of the different dike fillings show remarkable differences (Oskierski and Bischoff, 1983). The melt breccia dike Type 1-A is neither comparable to its host rock compositions nor to the overlying impact-melt sheets of the crater floor. The composition of the melt breccia Type 1-B is similar to the coherent impact melt. Type 2-A (polymict breccia) has a chemical composition quite similar to the surrounding country rock, whereas Type 2-B shows larger deviations from these; indicating that local material mainly contributed to the fillings of 2-A and less to the fillings of 2-B.

To obtain structural information about processes and directions of mass movement during the cavity modification stage, it is important to distinguish the fractures induced by the impact event from pre-impact target fractures, i.e. to find breccia dikes which fill impact induced fractures. Therefore, the spatial orientation of almost all breccia dikes was recorded. Fig. 14 shows the results represented in the lower hemisphere of the Schmidt net. Although data have been collected in a crater area which represents only about a quarter of the whole structure, the extension of the dikes is obviously randomly distributed with slight maxima of N-S striking dikes. Like the direction pattern of joints and fractures as described in par. 2.2, the dip angles show a great variety from flat lying

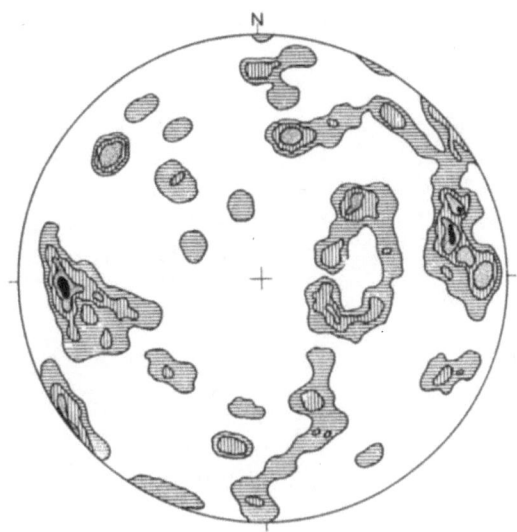

Fig. 14
Spacial orientation of breccia dikes,
poles to 94 contact planes;
contours at 2.4.6 and 9 per cent.

to vertical dikes. Sizes, frequencies and direction patterns of fractures in the crystalline basement of the crater floor vary in detail from exposure to exposure.

The spatial orientation of breccia dikes from the quarries of Champagnac (10), Laurière (3) and Pierre Folle (5) (Figs. 15—17) are rather similar to the orientation patterns of the tectonic fractures. Additionally, in the Laurière and Pierre Folle quarries a series of flat lying dikes exist which are dipping in various directions. The Champagnac maximum of N-S aligned dikes corresponds to a submaximum measured for tectonic fractures in this quarry (Fig. 1, diagram 10). As already mentioned, the orientation of tectonic elements is characterized by an extreme number of different directions, but is also displaying an apparent overall maximum of N-S striking fractures. Against this background, the measurements of breccia filled dikes indicate that not only primary fractures but also new systems of impact-induced fractures as presumed above (par. 2.2.1.1) were used during emplacement of breccia fillings.

6 Discussion

The results of the structural analysis in the Rochechouart basement, especially the greater variety of joints orientation and density, are in accordance with theoretical predictions (Curran et al., 1977) on the influence of an impact upon target rocks. On the other hand it could be shown that some of the impact energy did not lead to fracturing and brecciation but was absorbed by small movements along already existing fractures.

The coexistence of several peculiarities in the Champagnac quarry, such as low dipping shear fractures, pseudotachylite veins and hydrothermal-pneumatolytical alterations, possibly could indicate one and the same cause. In the following we will discuss two possibilities for the origin of the main fault planes A (Fig. 4) and the consequences upon the sequence of events in the crater floor.

1. The main fractures (A) could have been generated already at late Hercynian time. This hypothesis is favored due to the existence of many quartz veinlets (Qu-1) in the quar-

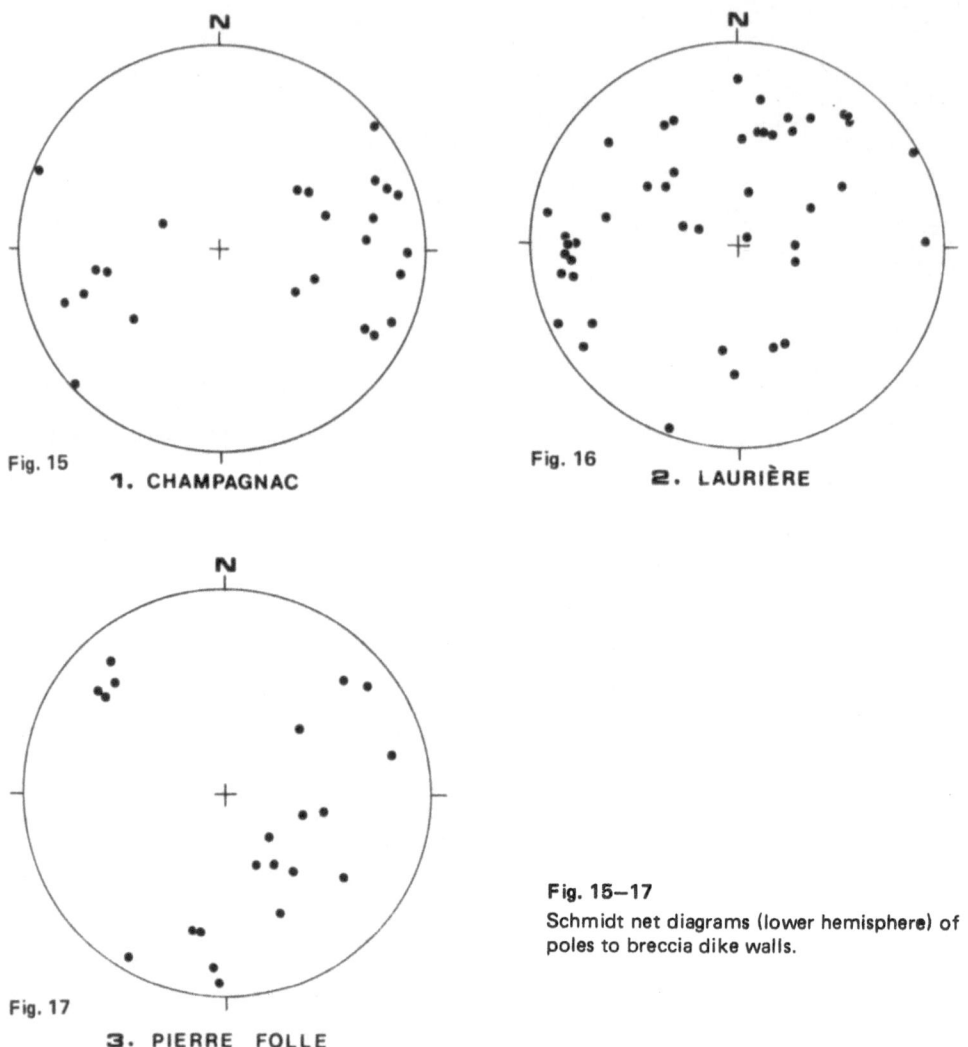

Fig. 15

1. CHAMPAGNAC

Fig. 16

2. LAURIÈRE

Fig. 17

3. PIERRE FOLLE

Fig. 15—17
Schmidt net diagrams (lower hemisphere) of
poles to breccia dike walls.

ry, which are clearly older than the breccia zones. As the breccias can be correlated with the impact event, the quartz veins must have been formed in pre-impact, most probably late Hercynian times. By the fact that many of the PT veins and the fault system A are roughly parallel to the Qu-1 veins, one could suppose that these too were originated at Hercynian time.

Other observations — the filling of the shear fractures with PT, quartz and sulfides, the sequence of brecciations, shear fractures and PT-quartz fillings as well as the Rb/Sr isotope data — clearly exclude a possible Hercynian age for these events. The relationship between the PT-quartz veins and the shear-fault system (A) makes it also unlikely that these faults were formed by Hercynian tectonics. Therefore, it is more probable that

2. The fault system (A) and the PT filling are both impact induced.

At Champagnac the large faults A_1 and A_2 (Fig. 4) are listric curved planes, as could be especially seen at an earlier stage of quarrying, when the lower part of the planes only had a dip of about 20°. Such fracture morphology is often found at a shallow level in the crust, in the form of thrusts under tangential stress as well as normal faults resulting from gravity gliding (land slides, crater wall failure ...). The sense of movement along these planes in the quarry is not clear, but the abundant auxiliary feather joints indicate down-throw movements. Supposedly, the PT veins following these fault planes were formed by frictional melting; the necessary heat demands rather a quick movement of the two rock segments under high pressure (Sibson 1975). During the impact at Rochechouart, the movements could have been caused by different impedances of the rocks on either side of the fault while under compression of the shock wave. But in the quarry there is no evidence for lithological and structural differences on each side of the fault planes, so that different impedances are unlikely. However, during the compressional stage of the impact event, movements might have taken place along some of the preexisting Qu-1 veins. This might explain the high frequency of secondary mobilized quartz together with PT-melt.

On the other hand if one ascribes the origin of the shear zone (A) to the modification stage of the impact crater, these fractures could have been formed at special sites of the originating crater, as is known from other impact structures (e.g. Sierra Madeira, Wilshire et al. 1972, Gosses Bluff, Milton et al. 1972). Relatively low-dipping shear planes are developed below the marginal depression between the crater rim and the central uplift of complex craters. Similar glide planes, but more steeply and radially dipping, are met in the outer zone of a central uplift. Dence et al. (1977) postulated their existence as a consequence of inward moving of large rock slabs into the region of a central uplift, caused by wall and floor failure of the transient cavity. From the situation of the Champagnac quarry within the Rochechouart crater and on the assumption that the rocks in the quarry were not rotated as a whole, i.e. that the orientation of the large fault plane still represents its original position, one could regard the listric curved planes for the glide surfaces upon which rock slabs moved under gravitation into the inner parts of the crater. It remains uncertain if the gravitational gliding of rock masses in the crater basement of Rochechouart would have been sufficient to produce frictional melts, or if, as supposed above, only rapid movement of blocks under the high pressure of the shock wave could have led to PT-melts. Thus if the second assumption is true, two different stages of movement along the shear fractures A must be distinguished: first, a thrust under compression to produce the PT-melt; later, a conversion of the sense of movement under gravitational glide.

Whatever may be true, the fault system A cannot be interpreted as part of movement planes within or at the border zone of a central uplift, because of its contrary orientation with a dip to the center. Lambert (1982) for several reasons doubts the existence of a central uplift in the Rochechouart crater floor, though other impact craters of the same diameter range in crystalline targets do show a central uplift. Oskierski (1983), who mapped in detail a part of the central region of the Rochechouart crater, points out that within an area of 3km^2 the allochthonous clastic and melt breccias are missing. That could mean that this zone represents the top of a small central peak, now eroded, or that the uplift basement was not primarily covered by breccias.

By gravity measurements in the Rochechouart structure, Pohl et al. (1978) discovered a minimum of the residual Bouguer anomaly in the order of 8 to 10 mgal in the crater center. They interpret this central anomaly as a zone extending several km under the present surface in which the crystalline rocks suffered a strong and cataclastic loosening.

A mass deficit like that could be related to the formation of a central uplift. As there are no clear structural proofs that the PT-melt was formed in situ on the fracture planes, one can doubt their origin by frictional melting. Instead the melt could have been injected from the large melt reservoir in the central region of the transient crater cavity. Then the PT-veins would represent a kind of very thin melt breccia-dikes (as dike breccia Type 1-A). But the chemical composition of the PT-veins at Champagnac is quite different from the much more homogeneous impact melt. Likewise the PT-melt shows affinities to the immediately surrounding rocks (Reimold et al. 1984 b). Therefore it seems unlikely that the PT veins represent melt injected over very long distances, but that they were rather generated nearby at shear fractures.

7 Conclusions (Genetic Model)

According to Stöffler (1977), structural changes of the impact crater basement can generally be assigned to various phases during the formation of the impact structure. As already mentioned above, Lambert (1981) discovered two breccia dike types of different age in the Rochechouart structure. In the following discussion, it is attempted to assign the total impact-induced phenomena found in the target rocks of the Rochechouart structure (such as fractures, pseudotachylites, breccia dikes, vein fillings) to the various stages of crater development.

I. Compression and Excavation Phase

a. Fracturing, formation of pseudotachylites
    The oldest undoubtedly impact-induced deformations found in the crater floor are brecciation, and minor slip movements on preexisting tectonic fractures (joints and faults as well as newly developed fractures of different orientations). These early processes originated from the passage of the shockwave through the compressed subcrater basement (compression phase). Along the faults subjected to high compression, significant frictional resistances had to be overcome before slip movements could take place, of which intensive brecciation and frictional melting (PT) give widespread evidence. Locally, Variscan moderately dipping quartz veins were transformed into mylonite and PT.

b. Breccia dikes
    As shown above, fragments of Type 1 melt breccia are found in the polymict breccia fillings of Type 2 and Type 3 dikes. Obviously Type 1-A represents the oldest breccia filling. The period between injection of Type 1-A melt breccia into the basement and its incorporation into younger dike generations was long enough to bracket the total crystallization time of the melt breccia. There is petrographical evidence that 1-A clasts in polymict breccias were transported in the crystalline state. Findings in Type 1-A — such as low clast content, lack of lithic fragments, a mean size of fragments smaller than in other dike types, flow structures as well as insignificant width — lead to the conclusion that 1-A melt was injected under very high energy conditions. Similar conditions have to be proposed for the formation of Type 2-A dikes. The petrographical evidence (par. 5) implies that 2-A dike breccias were dynamically injected rather than gravitatively deposited. Necessary pressures were only available during the compression and excavation phase of the crater formation. The obvious time gap between formation of 1-A and 2-A breccias can be explained by assuming that Type 1-A melt was injected at the end of the compression phase, whereas 2-A

was formed during the excavation phase. Mathematical models have been used to describe material flow during the excavation phase (Maxwell 1977, Croft 1980). Material is partially ballistically ejected but material from a cone-shaped part of the central crater region is not subjected to ballistic ejection; it moves towards the forming crater walls on horizontal or vertical paths directed sideways and downwards, respectively. Energetic conditions on these flow directions in the central crater areas are in agreement with the aforementioned requirements for the formations of 2-A veins. Opening fractures are provided during relaxation of the compressed basement. The existence of Type 2-A veins (clastic polymict breccias) in parautochthonous monomict breccias (e.g. in the Champagnac quarry) proves the relatively early development of subcrater fracturing.

II. Excavation and Modification Phase

Flat lying slip planes are found in the Champagnac quarry just below the crater floor surface. Therefore only small blocks of basement rocks (~50m) could have slipped. Friction between them could not have been sufficient to produce melt along the slip planes. It follows that PT-melt was already produced during the compression phase. During the following excavation and modification phase this melt might have served as lubricant, thus facilitating further slipping of basement blocks towards the crater interior.

Formation of intrusive veins of Type 3-A (monomict breccias) also occurred at the end of the excavation phase. Type 3-A textures are similar to those of Type 2-A and indicate a dynamic intrusive mechanism. However, the energy required was most probably lower than in the case of the formation of Type 2-A breccia dikes since only locally-derived clasts are found in 3-A. Therefore, long transportation distances can be discarded. Relative to the previously described crater formation phases, the modification process has to follow immediately after — possibly starting already during — the excavation phase. According to Grieve et al. (1977), impact melt formed in the central crater pit is forced into lateral flow during the formation of a central uplift in the early modification phase. This requires that the melt was still unconsolidated during the uplift. As the cooling period of granitic impact melt is estimated to be in the seconds-to-minutes range, it has to be concluded that the modification phase is set off very early in the crater history. During this phase, relaxation of the crater basement leads to continuous formation of fractures and veins. It can easily be comprehended that impact melt is gravitatively deposited in these fractures. Thus the existence of breccia Type 1-B impact melt texturally and chemically identical to the coherent impact melt (e.g. Babaudus or Chassenon) is explained. The petrographical characteristics of the polymict breccia Type 2-B require similar formation. This breccia variety is texturally comparable to the polymict fall-back breccias found above the crater floor. We believe that petrographical and stratigraphical evidence favor formation of Type 2-B polymict breccias by gravitative deposition of fall-back breccias into wide subcrater fractures (cracks) during the modification phase as predicted for melt breccia Type 1-B.

III. Postdeformational Modification

After completion of the described modification of the subcrater basement by block movements and dike formation, pneumatolytic and hydrothermal alterations took place along fault and brecciation zones. In many veins quartz, sulfides and carbonate crystallisations can be observed partially enclosing or intergrown with PT-melt. On the basis of petrographical, chemical and isotopic data, Reimold et al. (1984 a, b) suggested a four-stage genetic model for the formation of these vein fillings, characterized by close inter-action of pneumatolytic-hydrothermal phases and pseudotachylitic melt.

1. Lateral tectonic movements triggered by the impact of the Rochechourt meteorite (186 Ma ago) locally produce pseudotachylitic melt and clastic breccias of weak shock metamorphism.
2. Mechanical mixing of pseudotachylite melt with high-pressure fluid medium of inter-mediate temperature ($\sim$700°C).
3. Cooling period leading to hydrothermal crystallization, recrystallization, or partial recrystallization of pseudotachylitic meltrock or glass.
4. Formation of fine-grained carbonate veins crystallized from cold solutions infiltrated since the completion of the modification phase.

As already indicated in this model, there is clear evidence for crystallization from silicic-sulfidic-carbonaceous fluids of hot (300–700°C) as well as of relatively cool ($<$300°C) state. After this last period of mineralization, no considerable tectonic movements (slips $>$1 cm) or mineralogical changes occurred as indicated by the undeformed minerals of hydrothermal vein fillings.

Acknowledgements

Financial assistance from the Deutsche Forschungsgemeinschaft is gratefully acknowl-edged. The authors wish to thank Dr. W. U. Reimold for useful discussions. Special thanks are given to Mrs. Hartung, Mrs. McCormack and Dipl. Geol. A. Müller for technical assistance.

References

Chevenoy, M., 1958: Contribution à l'étude des schistes cristallins de la partie nord-ouest du Massif Central français. Mém. Serv. Carte Geol. Fr., Paris, 418 p.

Croft, S. K., 1980: Cratering flow fields: Implications for the excavation and transient expansion stages of crater formation. Proc. Lunar Planet. Sci. Conf. 11, 2347–2378.

Curran, D. R., D. A. Shockley, L. Seaman and M. Austin, 1977: Mechanism and models of cratering in earth media. In: Roddy, D. J., R. O. Pepin and R. B. Merrill (eds.): Impact and Explosion Crater-ing. Pergamon Press, New York, 1057–1087.

Currie, K. W., 1972: Geology and petrology of the Manicouagan resurgent caldera. Quebec. Geol. Surv. Can. Bull. 198, 153.

Dence, M. R., R. A. F. Grieve and P. B. Robertson, 1977: Terrestrial impact structures: Principle characteristics and energy considerations. In: Roddy, D. J., R. O. Pepin and R. B. Merrill (eds.): Impact and Explosion Cratering. Pergamon Press, New York, 247–275.

Gault, D. E., W. L. Quaide and V. R. Oberbeck, 1968: Impact cratering mechanics and structures. In: French, B. and N. M. Short (eds.): Shock Metamorphism of Natural Materials. Mono Book, Balti-more, 87–99.

Grieve, R. A. F., M. R. Dence and P. B. Robertson, 1977: Cratering processes: As interpreted from the occurrence of impact melts. In Roddy, D. J., R. O. Pepin and R. B. Merrill (eds.): Impact and Explosion Cratering. Pergamon Press, New York, 791–814.

Lambert, P., 1974 a: Etude géologique de la structure impactitique de Rochechouart (Limousin, France) et son contexte. Bull. BRGM Sec. I, 3, 153–164.

Lambert, P., 1974 b: La structure impactitique de Rochechouart (Limousin, France) et son contexte structural régional, par interpretation de "photosatellite", image ERTS. Bull. BRGM Sec. I, 177–183.

Lambert, P., 1977: The Rochechouart crater: Shock zoning study. Earth Planet. Sci. Lett. 35, 258–268.

Lambert, P., 1981: Breccia dikes: Geological constraints on the formation of complex craters. In: Multi-ring Basins., Proc. Lunar Planet. Sci. Conf. 12A, 56–78.

Lambert, P., 1982: Rochechouart: A flat crater from a clustered impact. Meteoritics 17, 240—241.

Maxwell, D. E., 1977: Simple Z model of cratering, ejection, and the overturned flap. In: Roddy, D. J., R. O. Pepin and R. B. Merrill (eds.): Impact and Explosion Cratering. Pergamon Press, New York, 1003—1008.

Melosh, H. J., 1977: Crater modification by gravity: A mechanical analysis of slumping. In: Roddy, D. J., R. O. Pepin and R. B. Merrill (eds.): Impact and Explosion Cratering. Pergamon Press, New York, 1245—1260.

Milton, D. J., B. C. Barlow, R. Brett, A. R. Brown, A. Y. Glikson, E. A. Manwaring, F. J. Moss, E. Č. E. Sedmik, J. Van Son and G. A. Young, 1972: Gosses Bluff impact structure, Australia. Science 175, 1199—1207.

Oskierski, W., 1983: Geologisch-petrographische Untersuchungen im Zentralbereich der Impaktstruktur von Rochechouart, SW Frankreich. Diploma Thesis, Univ. Münster.

Oskierski, W. and L. Bischoff, 1983: Petrographic, geochemical and structural studies on impact breccia dikes of the Rochechouart impact structure, SW France. Lunar and Planet. Science 14, 584—585, Houston.

Pohl, J., K. Ernstson and P. Lambert, 1978: Gravity measurements in the Rochechouart structure. Meteoritics 13, 601—604.

Raguin, E., 1972: Les impactites de Rochechouart (Haute Vienne), leur substratum crystallophyllien. Bull. BRGM Sec. I, 3, 1—8.

Reimold, W. U., J. Nieber-Reimold, W. Oskierski and A. Rehfeldt, 1983 a: A geochemical and geochronological study on amphibolites and granitic rocks from the Haut Limousin, Massif Central. Fortschr. Mineralogie, 61, 1, 178—180.

Reimold, W. U., W. Oskierski and A. Schmidt, 1983 b: Rb-Sr age dating of the Rochechouart impact event and geochemical implications for the formation of impact breccia dikes. Meteoritics 18, 385—386.

Reimold, W. U., L. Bischoff, W. Oskierski and H. Schäfer, 1984 a: Genesis of pseudotachylite veins in the basement of the Rochechouart impact crater, France. I. Geological and petrographical evidence. Lunar Planet. Science 15, 683—684.

Reimold, W. U., L. Bischoff, W. Oskierski, A. Rehfeldt and A. Schmidt, 1984 b: Genesis of pseudotachylite veins in the basement of the Rochechouart impact crater, France. II. Geochemical evidence and a genetic model. Lunar Planet. Science 15, 681—682.

Robertson, P. B. and R. A. F. Grieve, 1977: Shock attenuation at terrestrial impact structures. In: Roddy, D. J., R. O. Papin and R. B. Merrill, (eds.): Impact and Explosion Cratering. Pergamon Press, New York, 687—702.

Sibson, R. H., 1975: Generation of pseudotachylite by ancient seismic faulting. Geophys. J. R. astr. Soc. 43, 775—794.

Stöffler, D., 1971: Progressive metamorphism and brecciated crystalline rocks at impact crater. J. Geophys. Res. 76, 5541—5551.

Stöffler, D., 1977: Research drilling Nördlingen 1973: Polymict breccias, crater basement, and cratering model of the Ries impact structure. Geologica Bavaria 75, 443—458.

Stöffler, D., D. E. Gault, J. Wedekind and G. Polkowski, 1975: Experimental hypervelocity impact into quartz sand: Distribution and shock metamorphism of ejecta. J. Geophys. Res. 80, 4062—4077.

Wilshire, H. G., T. W. Offield, K. A. Howard and D. Cummings, 1972: Geology of the Sierra Madera cryptoexplosion structure, Pecos County, Texas. US Geol. Survey Prof. Paper, 599-H.

Impact Structures in Brazil

Alvaro Penteado Crosta

Universidade de Campinas, Instituto de Geociencias, Caixa Postal 6152, 13.100 Campinas, SP. Brazil
Present address: A/C Dr. John McM. Moore, Imperial College/RSM, Geophysics Section 29 Prince
Consort Road, London SW7 2BP, UK

Key Words

Impact structure
Crater
Brazil

Abstract

Studies carried out in the last ten years resulted in the recognition of several circular
structures in the Brazilian territory which have, as common characteristics, the presence
of similarities with well-known structures in other countries and planets attributed to
impact with celestial bodies, and the absence of evidence for a terrestrial endogeneous
origin. Those structures presently number six, ranging in diameter from three to 40 kilometers. Two of them, Araguainha and Serra da Cangalha, were studied in some detail
and presented enough evidence for their impact origin, including impact metamorphism
features and morphologic/structural styles. The other four, Colonia, São Miguel do
Tapuio, Vargeão and Riachão, are still poorly known, but the available data are very
suggestive for their impact origin.
Similar structures were identified by remote sensing studies, mainly in the Amazon
region, but no field data are available yet. It is believed that the whole number of impact
structures in Brazil is more than six, mainly due to the size of the territory (8.5 million
square kilometers).

Introduction

The main purpose of this paper is to provide a synthesis of the knowledge about the
Brazilian impact structures. It includes the results of studies carried out in the last years
by many authors, which allowed the recognition of several circular structures presenting
similarities with well known impact structures on Earth and other bodies of the Solar
System.
It is believed by the author that the whole number of impact structures in the 8.5 million
square kilometers of Brazilian territory is higher than the six described here. Therefore,
the present small number is due to the absence of surveys for impact structures detection, the low degree of geological knowledge about the huge territory, which is mapped
only in small scales (1 : 1 000 000 and 1 : 500 000), as well as to the dense vegetation cover,

mainly in the Amazon Region (1/5 of the territory), which makes difficult the detection of circular anomalies by remote sensing techniques.

On the terrestrial cratering record, by Grieve and Robertson (1979), only four Brazilian structures appear: Araguainha Dome, Serra da Cangalha, Colonia and São Miguel do Tapuio. Araguainha Dome and Serra da Cangalha were classified as probable impact craters and Colonia and São Miguel do Tapuio as possible impact craters. Now, two other structures can be added to the probable impact craters list: Riachão and Vargeão Dome. Other structures have been pointed out recently as possible impact craters in the Amazon Region and Minas Gerais State, but no field data are available yet.

A synthesis of the available data for the six structures will be given next.

Araguainha Dome

Araguainha Dome is a 40-kilometer diameter circular structure in paleozoic sediments of the Parana Basin. The center is located $16°47'S$ and $52°59'W$ and the structure is divided in the middle by the Araguaia River, which is the boundary between Goias and Mato Grosso states.

Regional geological surveys (Northfleet et al., 1969; Silveira Filho and Ribeiro, 1971) described the structure as the result of an alkalic intrusion which arched the paleozoic sediments into a dome shape. Those authors associated this intrusion with the widespread basaltic/alkalic volcanism of the Jurassic/Cretaceous of the Parana Basin.

With the work of Dietz and French (1973) and Dietz et al. (1973), it was defined as the larger and better known Brazilian impact structure. Using satellite ERTS (actually LAND-SAT) images (Fig. 1), they located the Araguainha Dome and associated its shape with other astroblemes of the northern hemisphere. The thin-section analyses of the polymict

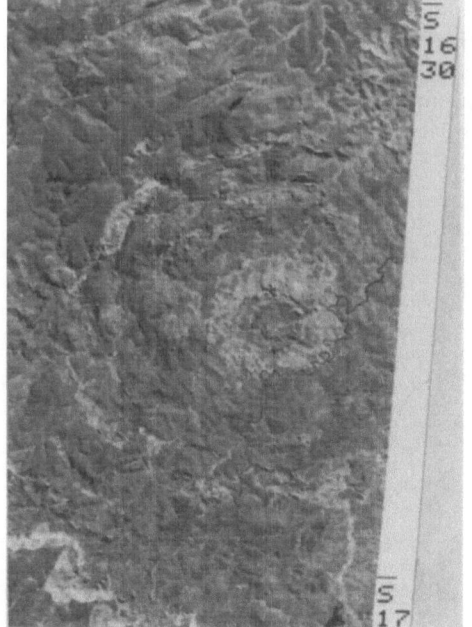

Fig. 1
MSS/LANDSAT band 7 image showing the Araguainha Dome. Lower image width 36 km.

breccias found in the center of the dome by those authors showed planar features (shock lamellae) in quartz grains. Those features, as well as the occurrence of breccias containing rock fragments showing shock deformation features and melting, were related by the authors to an impact phenomenon of a celestial body (comet?) upon the Earth's surface. Later studies done by Crosta et al. (1981), Theilen-Willige (1981) and Crosta (1982) extend this evidence through the analysis of morphologic, structural and petrographic features. Those studies showed a whole information set, totally in accordance with the criterion established by Khryanina (1979) on the identification of impact structures.

In the core, the Araguainha Dome has a circular central uplift 6,5 kilometers in diameter (Fig. 2) and is surrounded by an uplifted outer rim. In this outer rim, semicircular grabens formed by ring faults comprise strongly deformed Permian sediments, while in the dome interior it is possible to note an alternation of successive depressed and uplifted zones, showing in some places strongly folded and faulted portions. Those morphologic-structural features give the dome a multicircular aspect, easily visible in MSS/LANDSAT images.

In addition to the breccias with vitreous matrix, those studies identified several impact metamorphism features, mainly in the granitic basement exposed in the middle of the central uplift and in the rock fragments of the polymict breccias. Those features include the occurrence of shatter cones in sandstones, planar features in feldspar and micas besides the planar features in quartz, the passage of crystalline to amorphous state in minerals without melting, in which the vitreous phase retains the textural and morphologic characteristics of the crystalline phase, kink bands in feldspars and micas, mechanic deformations and ruptures restricted to the mineral grains, general oxidation of iron-magnesian minerals and several other features.

Crosta (1982) presents the results of radiometric age determination (K-Ar method) done in a sample of the granitic basement which outcrops in the middle of the central uplift. This determination gave values of 362.6 ± 13.2 Ma to the mafic fraction (biotite) and 283.6 ± 17.2 Ma to the felsic fraction (feldspar). The author's interpretation explains those results as intermediary ages between the last thermo-metamorphic event (Brasiliano Cicle — 650 Ma) and the age of the impact, since the mineralogical analysis of the sample dated showed only light deformations by shock, which indicates that there was not sufficient energy to remove entirely the pre-existent argon from the rock. On the other hand, stratigraphic observations done by the same author give to the structure an age between the transition Triassic-Jurassic and the Upper to Medium Permian.

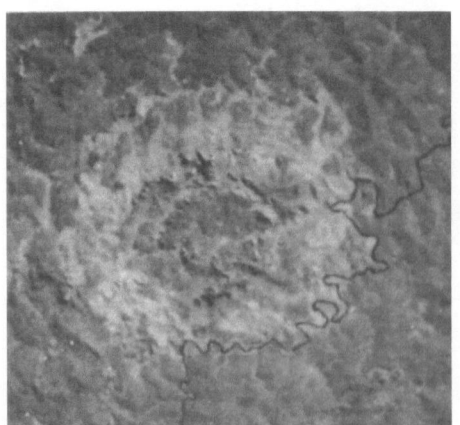

Fig. 2
Portion of MSS/LANDSAT image showing Araguainha Dome's central uplift. Image width 22 km.

Serra da Cangalha

The Serra da Cangalha structure (Fig. 3) is located at 8° 05'S and 46° 52'W, in the northern limit of the Goias State. It is a 12-kilometer diameter circular structure in paleozoic sediments in the Parnaiba Basin.

The authors who first proposed the impact origin for this structure were Dietz and Franch (1973). This proposition was based on the circular shape of the structure with a 3-kilometer diameter central uplift composed of a circular mountain range 250 meters high in Poti sandstones (Permian/Carboniferous) with an exposition of the lower sequence (Longa shales) in the center, and on the abscence of volcanic activity evidence in the drillings made at the center and on the impossibility of salt diapirism in the lower clastic sequences.

Later studies done by McHone (1979) and Santos and McHone (1979) added new data to those. Among them are the discovery of shatter cones in the central uplift's interior, in a conglomerate at the bottom of Poti Formation. The conglomerate's pebbles show a dense fracture system with millimetric width and the quartz grains are intensively fractured. McHone (1979) also presents some electron microscopy results which show microspherules constituted by melt material, injected into microfractures.

Fig. 3
MSS/LANDSAT image of the Serra de Cangalha structure. Image width 12 km.

Colonia

The Colonia structure (Fig. 4) has its center located at 23° 48'S and 46° 42'W, very close to São Paulo city, in São Paulo State.

It is a perfect circular depressed structure, with a diameter of 3 kilometers. The structure is in crystalline rocks of the pre-Cambrian basement and is full of Quaternary sediments.

Fig. 4
Aerial photograph of Colonia structure

Studies by Kollert et al. (1961), including geological and geophysical surveys, showed the 350 meters basement depth in the center of the structure and its bowl-shaped configuration.

The structure seems to have been formed in Tertiary or Quaternary periods, since the preservation of the outer rim and the degree of erosion are satisfactory.

In Colonia the identification of impact features becomes more difficult due to the sedimentary fill. On the other hand, some analogy is possible between the dimension of the structure and the impact craters less than 3.8 kilometers in diameter. For this category, Grieve and Robertson (1979) present one equation to associate the diameter to the depth of the crater. For craters formed in crystalline rocks with good morphologic preservation degree, the depth (B) of the crater is equal to the diameter (D) elevated to 0.829 and multiplied by 0.159. Using this relation to the 2625 meters of diameter of Colonia structure, the result is 354 meters of depth, perfectly in accord with the depth obtained by Kollert et al. (1961) through geophysical methods.

São Miguel do Tapuio

This structure is located at 5° 38'S and 41° 24'W, in the northeastern portion of the Piaui State, and has around 20 kilometers diameter in the dominion of sedimentary rocks of the Parnaiba Basin (Cabeças Formation).

The circular structure of São Miguel do Tapuio (Fig. 5) has been cited by many authors as the product of an igneous intrusion (laccolith) not yet outcropping (Siqueira Filho, 1970; Nunes et al., 1973; Lima, 1978). Recently, Torquato (1981) concluded through geological studies about a probable impact origin. This conclusion is based on some evidence on its morphologic style, the transformation of the Cabeças sandstone into a quartzite at the center of the structure, the asymmetry of its scarps which are abrupt and elevated in the western part and soft in the southeastern, showing the probable direction of the collision which would have reached the surface with a certain angle, and the drop of 50 meters between the scarp borders and the structure's center.

Torquato (1981) suspects a pre-opening of the Atlantic Ocean age for this structure, because it is affected by faults of that age.

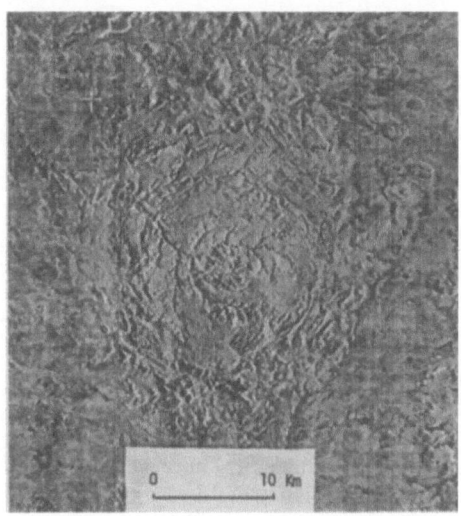

Fig. 5
Radar image of São Miguel do Tapuio.

Since there occur many basic intrusions (sills and laccoliths) with variable thickness in the region as well as a lack of sufficient evidence of impact metamorphism, it is premature to say São Miguel do Tapuio is an astrobleme, although the evidence found until now is very promising.

Riachão

The Riachão structure (Fig. 6) is located at 7° 46′ S and 46° 39′ W, in the southern portion of Maranhão State, and it has a 4-kilometer diameter in sedimentary rocks of the Parnaiba Basin. Riachão is only at 45 kilometers distance from Serra da Cangalha structure, in the northeastern direction. It was first discovered by the members of an Apollo mission in 1975. The name Riachão was proposed by the NASA researchers.

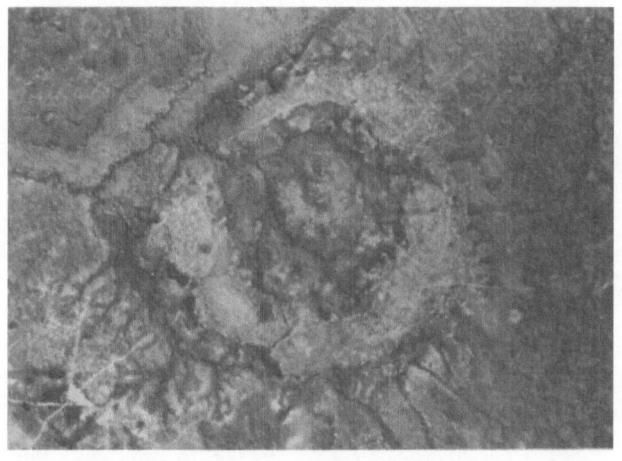

Fig. 6
Aerial photograph of the Riachão structure.
Image width 8 km.

McHone (1979) and Santos and McHone (1979) presented the results of a visit to this structure, which apparently involves only sandstones from the Pedra de Fogo Formation. Those authors have identified some evidence of an impact phenomena, such as the existence of a central uplift, the occurrence of polymict breccias inside the structure which were not detected in any of the stratigraphic wells done for hydrocarbon prospection in the region, as well as the uncommon pattern of microfractures in quartz grains observed in thin sections of polymict breccias. But this evidence is not conclusive for an impact metamorphism in Riachão.

Vargeão Dome

The Vargeão Dome (Fig. 7) is located at $26°50'S$ and $52°07'W$ and is 12 kilometers in diameter. The structure is located in igneous and sedimentary rocks from the São Bento Group of Parana Basin.

The discovery of this circular structure is due to Paiva Filho et al. (1978) who, through geological photointerpretation, had detected a radial and annular fracture pattern which limits the topographic depression found in the area, exhibiting drops up to 150 meters in relation to the surrounding surface. These authors had also distinguished the occurrence, at the center of the structure, of sandstones with eolian cross-bedding, and related them with the Botucatu Formation forming therefore a stratigraphic window.

Barbour Jr. and Correia (1981) defined these sandstones as conglomeratic with clay intercalations, showing dips in the order of 40 degrees. They detached the existence of a normal fault system with direction $N60-70°E$ which crosses the structure and extends itself regionally and of an annular and radial fault system also with a normal character, which limits the dome, as well as the intense fracturing and jointing of the rocks in the structure's interior. Breccia zones were also observed by the authors in many places. Among the hypotheses related by Barbour Jr. and Correia (1981) to explain the origin of the structure are the depression of blocks by normal faults, meteoritic impact, escape of volcanic gases and deep igneous intrusion.

Fig 7
Radar image of Vargeão Dome.
Image width 20 km.

Paiva Filho et al. (1982) discuss some ideas about the origin of the Vargeão Dome point-ing out that, since there is no evidence of a volcanic extrusive or intrusive activity and as the geophysical data available do not denote the presence of a deep intrusion, the most reasonable hypothesis is the impact of a celestial body. The morphologic and structural similarities observed between the Vargeão Dome and many other abstroblemes, among which stands out the existence of a central uplift confirmed by the occurrence of Botu-catu sandstones at least 500 meters above its normal stratigraphic position, reinforce this hypothesis.

Recent studies of thin sections of the outcropping sandstone of the center of Vargeão Dome showed planar features in quartz grains in at least two different directions (Coutin-ho, 1983 — personal communication to the author).

Final Considerations

In the present level of geological knowledge of circular structures presented up to now in Brazil, whose origins are supposedly related to impact phenomena, only two displayed features which can be related for sure to the occurrence of impact. To the other four, however, evidence has been revealed which points to the supposition of this type of origin. This evidence is, in most cases, of morphologic and structural character.

On the other hand, it should yet be said that only Araguainha, Serra da Cangalha and Riachão were objects of specific studies aiming to identify impact metamorphism evi-dence, and from them only the last one had not presented safe results, although there have been revealed interesting features (microfractures). Then, it should be expected that other data will come to light as new research will be developed in those areas.

References

Barbour Jr., E. and W. A. G. Correia, 1981: Geologia da estrutura de Vargeão-SC. São Paulo, Pauli-petro. Relatorio RT-023/81, 22p.

Crosta, A. P., 1982: Mapeamento geologico do Domo de Araguainha utilizando tecnicas de sensoria-mento remoto. Master dissertation, Instituto de Pesquisas Espaciais (INPE), São Jose dos Campos, 108p.

Crosta, A. P., J. C. Gaspar and M. A. F. Candia, 1981: Feições de metamorfismo de impacto no Domo de Araguainha. Rev. Bras. Geoc., 11, 139—146.

Dietz, R. S. and B. M. French, 1981: Two probable astroblemes in Brazil. Nature 244, 561—562.

Dietz, R. S., B. M. French and M. A. M. Oliveira, 1973: Araguainha Dome and Serra da Cangalha, probable astroblemes. Congresso Brasileiro de Geologia, 27°.Aracaju. Bul. 1, 102.

Grieve, R. A. F. and P. B. Robertson, 1979: The terrestrial cratering record. Icarus 38, 212—229.

Khryanina, L. P., 1979: Meteoritic craters as geologic structures. Geotectonics 13, 246—254.

Kollert, R., A. Björnberg and A. Davino, 1961: Estudos preliminares de uma depressão circular na região de Colonia, São Paulo. Bul. of Soc. Bras. Geol. 10, 57—77.

Lima, M. I. C., 1978: Potencialidades das imagens de radar em mapeamentos geologicos. Congresso Brasileiro de Geologia, 30°. Recife, Vol. 1, 164—178.

McHone, J. F., 1979: Riachão Ring, Brazil: a possible meteorite crater discovered by the Apollo astronauts. Apollo-Soiuz Test Project, Summ. Sci. Rep., Vol. 11, 193—202.

Northfleet, A. A., R. A. Medeiros and H. Muhlmann, 1969: Reavaliação dos dados geologicos da Bacia do Parana. Petrobras Tech. Bull. 12, 291—346.

Nunes, A. B., R. F. F. Lima and C. N. B. Filho, 1973: Geologia da Folha SB-23 (Teresina) e parte da Folha SB-24 (Jaguaribe). Radam Project, Vol. 2.

Paiva Filho, A., C. A. V. Andrade and L. F. Scheibe, 1979: Uma janela estratigrafica no oeste de Santa Catarina: o Domo de Vargeão. Congresso Brasileiro de Geologia, 30°. Recife, Vol. 1, 408—412.

Paiva Filho, A., A. P. Crosta and G. Amaral, 1982: Utilização de dados de sensoriamento remoto no estudo estratigrafico e estrutural da Formação Serra Geral (Sul do Brasil). Simposio Brasileiro de Sensoriamento Remoto, 2°. Brasilia.

Santos, U. P. and J. F. McHone, 1979: Field report on Serra da Cangalha and Riachão circular features. Instituto de Pesquisas Espaciais (INPE), relat. 1458-NTE/153, 13p.

Silveira Filho, N. C. and C. L. Ribeiro, 1971: Informações geologicas preliminares sobre a estrutura vulcanica de Araguainha, MT. Goiania, DNPM relat.

Siqueira Filho, J., 1970: Geologia da Folha de Castelo do Piaui. Sudene, Recife. Geologia Regional Serie, n°. 15, 64p.

Theilen-Willige, B., 1981: The Araguainha impact structure, central Brazil. Rev. Bras. Geoc. 11, 91–97.

Torquato, J. R., 1981: O astroblema de São Miguel do Tapuio (PI). Ciencias da Terra 1, 37.

The Sudbury Structure, Ontario, Canada — A Review

B.O. Dressler*
Ontario Geological Survey, 903-77 Grenville St., Toronto, Ontario, Canada M7A 1W4

G.G. Morrison/W.V. Peredery*
Inco Limited, Copper Cliff, Ontario, Canada P0M 1N0

B.V. Rao
Dept. of Geology, University of Toronto, Toronto, Ontario, Canada M5S 1A1

Abstract

The Sudbury impact structure is characterized by ring fractures in the footwall rocks, by an overturned crater collar and by shock metamorphosed and brecciated footwall rocks. The Whitewater Group occurs within the Sudbury Basin and consists of the diaplectic, heterolithic, glass-rich breccias and melt bodies of the Onaping Formation, the mudstones of the Onwatin Formation and the wackes of the Chelmsford Formation. The Sudbury Igneous Complex intruded along the base of the Whitewater Group and is composed of norite, quartz gabbro and granophyre of the "Main Mass" and of the gabbroic, inclusion-rich rocks of the Sublayer. Chemical evidence suggests that the Complex was formed by impact-initiated remelting of a mafic source body. This mafic body is located at depth below the Complex and may have also acted as a source magma chamber for the older Nipissing gabbro.
An impact model is illustrated and a comparison with the Ries impact crater in Germany is presented. As well, some evidence forwarded by opponents of an impact origin of the Sudbury Structure is described.

1. Introduction

The Sudbury Structure is of Early Proterozoic age and straddles the present boundary of two major geological subdivisions of the Canadian Shield: the Superior Province (Archean), and Southern Province (Proterozoic) (Fig. 1). Metamorphic and igneous rocks of the Superior Province are unconformably overlain by Early Proterozoic metavolcanic and metasedimentary rocks of the Huronian Supergroup. The Creighton granitic pluton (2,333 + 33/−22 Ma, U-Pb zircon; Frarey et al., 1982), Murray granitic pluton (2388 + 20/−13 Ma, U-Pb zircon; Krogh et al., 1984) and Skead granitic pluton intrude the

---

* This paper is published with the permission of Inco Limited and of the Director, Ontario Geological Survey.

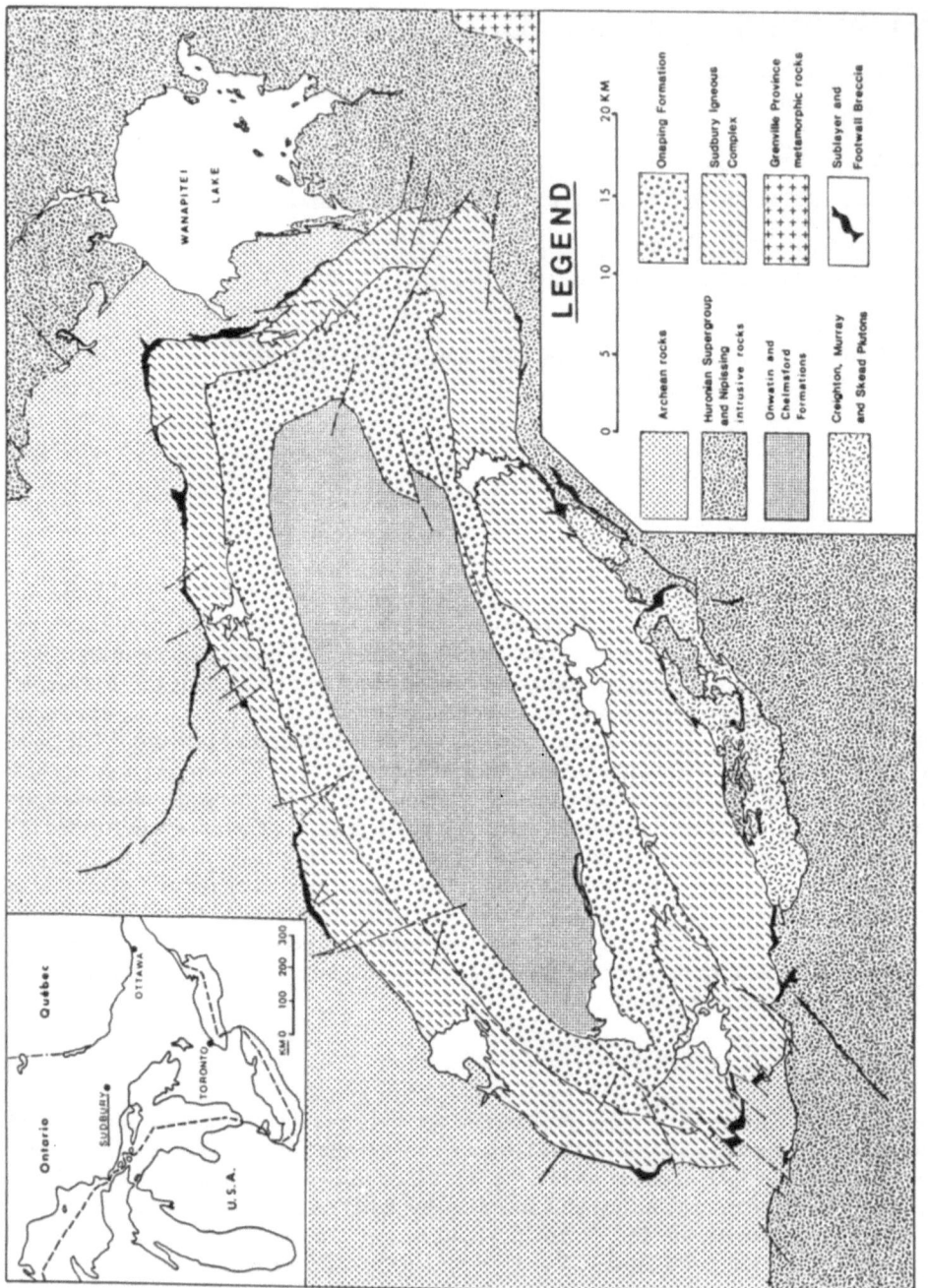

Fig. 1
General Geology of the Sudbury Structure.

Huronian sequence south and east of the Sudbury Igneous Complex. The Nipissing gabbro bodies ($2,150 \pm 50$ Ma, Rb-Sr, Van Schmus, 1965; Fairbairn et al., 1969) are widespread in the area.

Contained in the depression within the Sudbury Structure is the Whitewater Group, consisting of the Onaping, Onwatin and Chelmsford Formations. The Sudbury Igneous Complex occurs structurally below the Whitewater Group. Associated with the Sudbury Structure are pseudotachylites which cut the footwall rocks but not the Sudbury Igneous Complex or the Whitewater Group. Middle Proterozoic olivine diabase dikes ($1,250 \pm 50$ Ma, Van Schmus, 1965; Fahrig and Wanless, 1963) intrude all the foregoing rocks.

Southeast of the Sudbury Structure Middle to Early Proterozoic, high rank metamorphic rocks of the Grenville Province occur. The boundary of this province with the Southern Province is marked by the northeasterly trending Grenville Front Tectonic Zone (Lumbers, 1975).

Shock metamorphic features in the footwall rocks and in the Onaping Formation and the brecciation of the footwall rocks are related to the Sudbury Structure, interpreted in this paper as a meteorite impact crater.

The earth's largest nickel-copper sulphide deposits are associated with Sudbury Igneous Complex which was intruded into the Sudbury Structure. This consists of the Lower Zone norite, the Middle Zone quartz gabbro and the Upper Zone granophyres of the "Main Mass" and of several intrusive phases of the Sublayer. The Sublayer occurs at the lower contact of the Main Mass and also as dikes, commonly known as "offsets", in the footwall rocks.

The origin of the Sudbury Structure is still a matter of debate. In 1964 R. Dietz published his classic paper "Sudbury Structure as an Astrobleme" and since then many papers both in favour (Dence, 1972; French, 1972; Peredery, 1972a, b; Peredery and Morrison, 1984), and against an impact origin (Card and Hutchinson, 1972; Stevenson, 1972; Stevenson and Stevenson, 1980; Muir, 1984) have been published. The authors of this paper favour an impact origin for the Sudbury Structure. We are aware, however, that many problems of Sudbury geology remain unsolved and a final answer to all these problems is far from being reached. For more information the reader is referred to an Ontario Geological Survey publication "The Geology and Ore Deposits of the Sudbury Structure" (Pye, Naldrett and Giblin, 1984)

The present paper is subdivided in four parts. The first deals with impact related tectonic features and with shock features in the footwall rocks of the Sudbury Structure. A second part describes relevant features of the melt bodies and breccias of the Onaping Formation. The third part deals with the Sudbury Igneous Complex believed by the authors to be only indirectly related to the Sudbury Structure. In a last fourth part features of Sudbury Geology are described that are not easily accommodated in an impact scenario.

2. The Effects of Sudbury Impact of the Footwall Rocks of the Sudbury Structure

Ring Fractures and Overturned Crater Collar

On Landsat satellite images, a distinct ring fracture approximately 18 to 27 km distant from the Sudbury Igneous Complex can be recognized (Fig. 2). It is crudely concentric with the North Range Sudbury Igneous Complex, coincides with several outliers of metasediments of the Huronian Supergroup, and is similar in appearance to ring fault zones

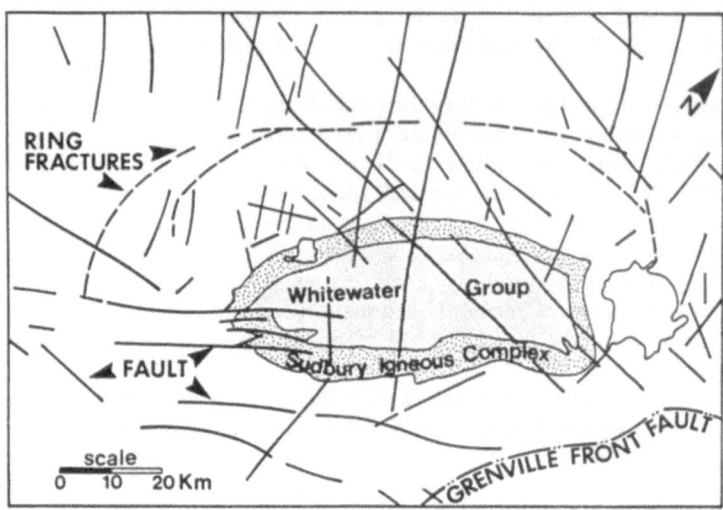

Fig. 2
Ring fractures north of the Sudbury Igneous Complex (as seen on Landsat images).

around some terrestrial and lunar craters. Ring and multi-ring systems are common on the earth's moon and other bodies in our solar system (Taylor, 1982).

Field observations of the ring fracture indicate the presence of vertical fault planes with vertically plunging slickensides. The strike of these faults is subparallel to the strike of the ring fracture zone as seen on the satellite image.

The South Range crater collar consists mainly of rocks of the Huronian Supergroup and granitic rocks of the Creighton and Murray Plutons. In general the Huronian rocks dip vertically or may be overturned. Farther away from the Complex, dip directions are away from the Complex and decrease in magnitude, from 90° to 35° south over a distance of about 2 km. This steepening of dip, including overturning of stratified rocks in the vicinity of the crater, has been observed at several other impact craters. Examples are the Vredefort Structure in South Africa (Daly, 1947), the Decaturville Structure in Missouri (Offield and Pohn, 1977), and the Meteor Crater in Arizona (Shoemaker, 1963).

Shock Metamorphic Features

In the Sudbury footwall[1] rocks, macroscopic and microscopic shock metamorphic features are common. South of the Sudbury Igenous Complex, however, microscopic features are largely destroyed due to regional metamorphic recrystallization.

Shatter cones occur around the entire Sudbury Igneous Complex for distances as much as 17 km away from it. Fig. 3 illustrates the distribution of shatter cones and includes data collected by Guy Bray and Geological Staff (1966) and 60 locations added in this paper. Many of the shatter cone apices point toward the central part of the Sudbury

---

1 The term "footwall" is defined as comprising rock units that are characterized by deformational and metamorphic features related to the Sudbury impact.

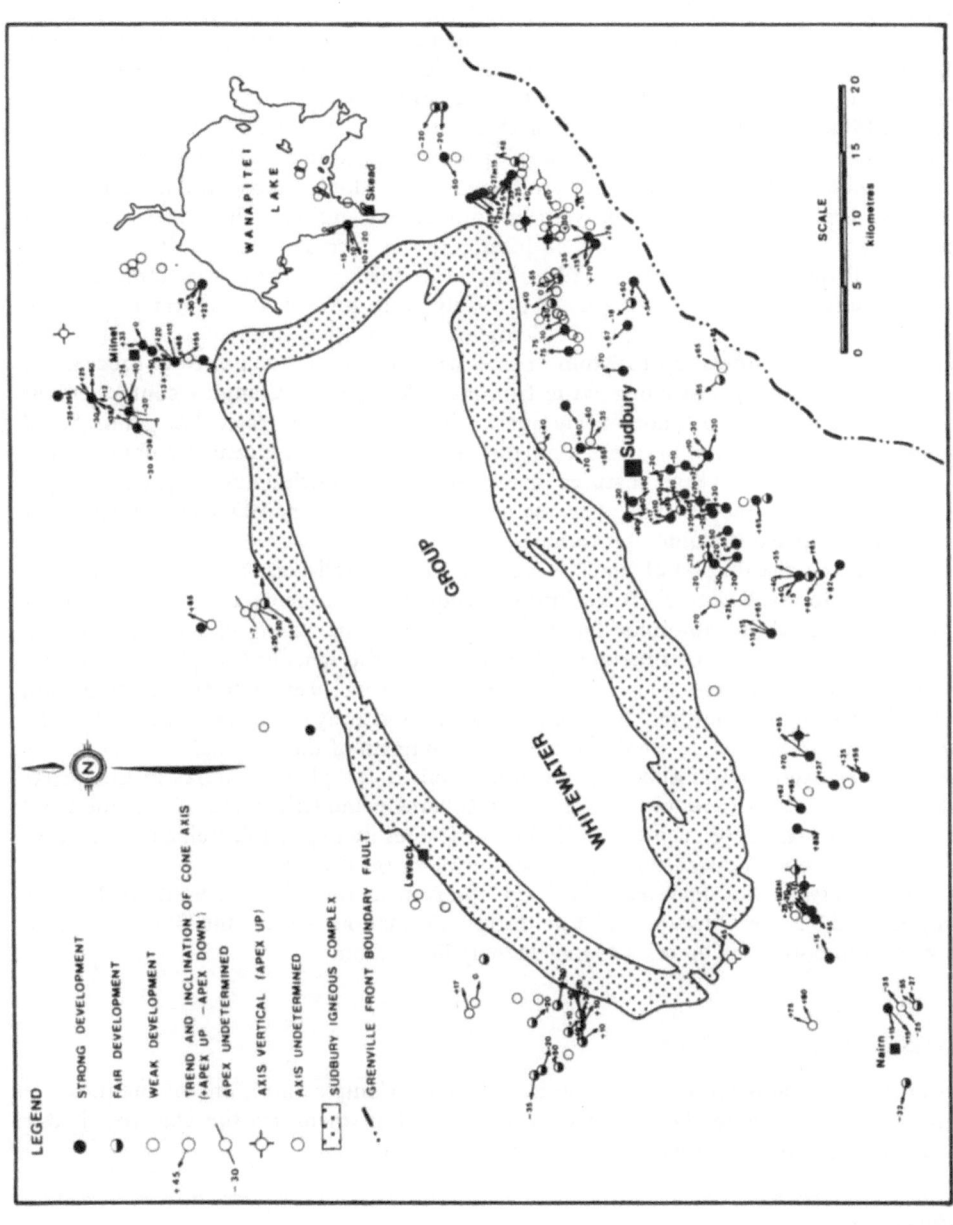

Fig. 3

Shatter cone location and orientations around the Sudbury Igneous Complex (from Dressler, 1984, after Guy-Bray and geological staff, 1966).

Basin, but random orientations at one and the same shatter cone location are also common. The cones are absent from rocks of the Sudbury Igneous Complex and Whitewater Group but have been observed in country rock fragments in the Onaping Formation (Dietz, 1972, Peredery, 1972b).

Kink bands in biotite and other micaceous minerals are the first sign of shock-induced mineral deformation seen under the microscope. Planar features in quartz, plagioclase and potassic feldspars have been observed in the footwall rocks and have been described by Dressler (1984). Partial or complete diaplectic isotropization was not noted in any rock forming mineral. Due to contact metamorphic recrystallization near the Sudbury Igneous Complex shock metamorphic features are commonly no longer recognizable. In the contact metamorphic aureole, however, very fine mosaic-like and spherulitic recrystallization patterns have been noted and could be interpreted as recrystallization after diaplectic feldspar and quartz glass. Dressler (1984) discusses further evidence that indicates also that diaplectic, isotropic mineral phases probably were present at the crater rim prior to the intrusion of the Sudbury Igneous Complex.

Filled vesicle-like features, up to 2 mm in size, were noticed in the Footwall Breccia and in an autochthonous gneissic migmatite from near the Igneous Complex contact. These consist of xenomorphic to idiomorphic quartz rimmed by small plagioclase grains. Small amount of carbonate may be present in the centres, and in one case, a stubby apatite crystal was observed. These features may represent a shock-induced quartz-feldspar crystallized melt trapped in an otherwise solid rock (G. Graup, Max-Planck-Institut für Chemie, Mainz, personal communication, 1983).

To illustrate the distribution of shock features in the footwall rocks a small section of the northwestern footwall (Levack Township) was chosen for a more detailed investigation of the zoning of shock metamorphic features (Fig. 4). A crude zoning of such features can be recognized. In the contact metamorphic aureole of the Igneous Complex the features are present only as relicts. Planar features in quartz have been observed up to 5.5 km north of the Complex, but may be present as far as 8 km away as shown by Dence (1972). Planar features in plagioclase were noted up to 3 km north of the Complex.

The shock pressures responsible for the formation of the planar features in the quartz grains were estimated for a few samples using Robertson and Grieve's (1977) method and are shown at the appropriate locations in Fig. 4. The result, in general, indicates a decrease in shock pressure with distance from the Sudbury Igneous Complex.

No shock metamorphic features have been observed in rocks of the Sudbury Igneous Complex and of the Onwatin and Chelmsford Formations within the Sudbury Basin. They are common in the breccias of the Onaping Formation.

Breccias in the Footwall Rocks

The breccias in the footwall of the Sudbury Igneous Complex are some of the most intriguing rocks related to the Sudbury Structure. Within them lies the clue to a better understanding of processes related to large impact craters.

Sudbury Breccias

Peripheral to the Sudbury Igneous Complex the footwall rocks host pseudotachylite breccia bodies known as Sudbury Breccia, previously also called Sudbury-type breccia, Levack Breccia or Common Sudbury Breccia (Fairbairn and Robson, 1942, 1943; Speers, 1957; Card, 1978; Dupuis et al., 1982, Dressler, 1984). The breccia occurs as nume-

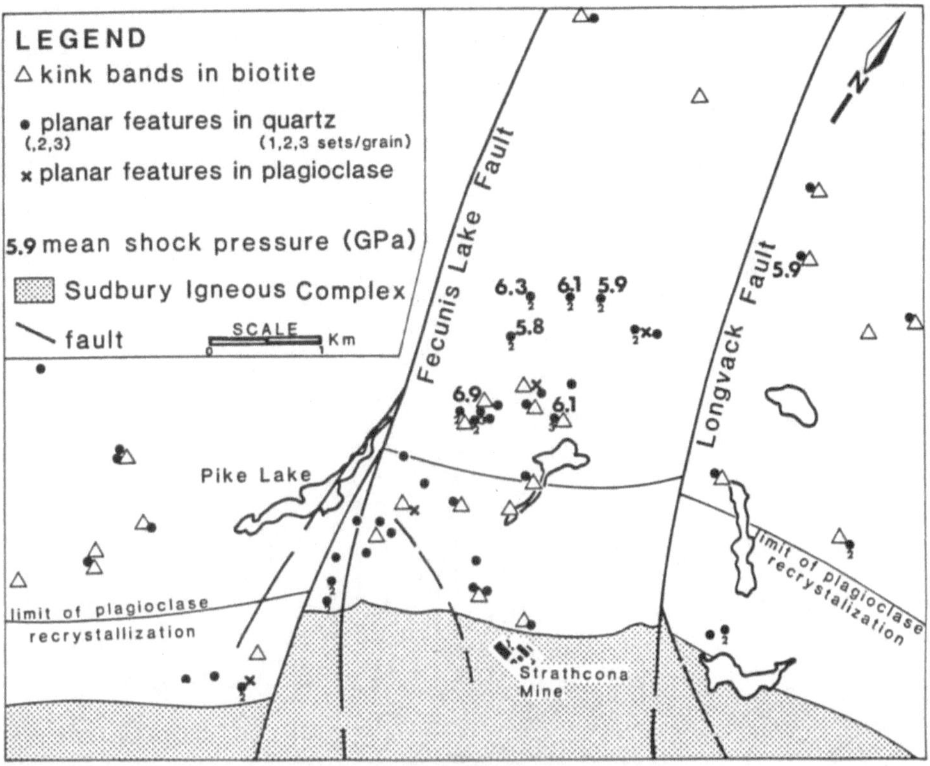

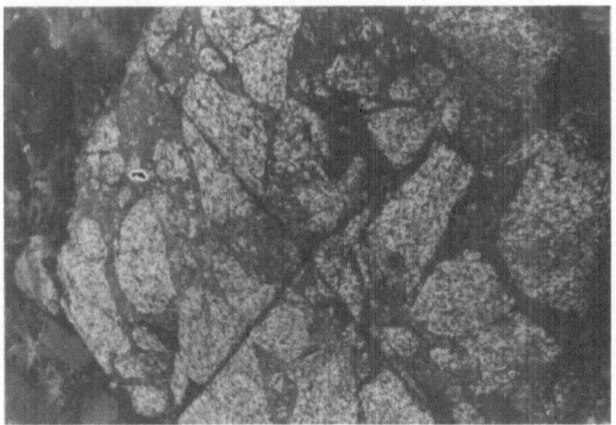

LEGEND

△ kink bands in biotite

• planar features in quartz
(,2,3)   (1,2,3 sets/grain)
× planar features in plagioclase

5.9 mean shock pressure (GPa)

▨ Sudbury Igneous Complex

∖ fault   SCALE   Km

Fecunis Lake Fault

Longvack Fault

6.3   6.1   5.9

5.8

6.9   6.1

Pike Lake

limit of plagioclase recrystalization

limit of plagioclase recrystallization

Strathcona Mine

Fig. 4
Zoning of shock metamorphic features in Levack Township, north of Sudbury Igneous Complex
(after Dressler, 1984).

Fig. 5
Sudbury Breccia in Creighton Granite. South of Sudbury Igneous
Complex. (Diameter of lens cap is 5 cm).

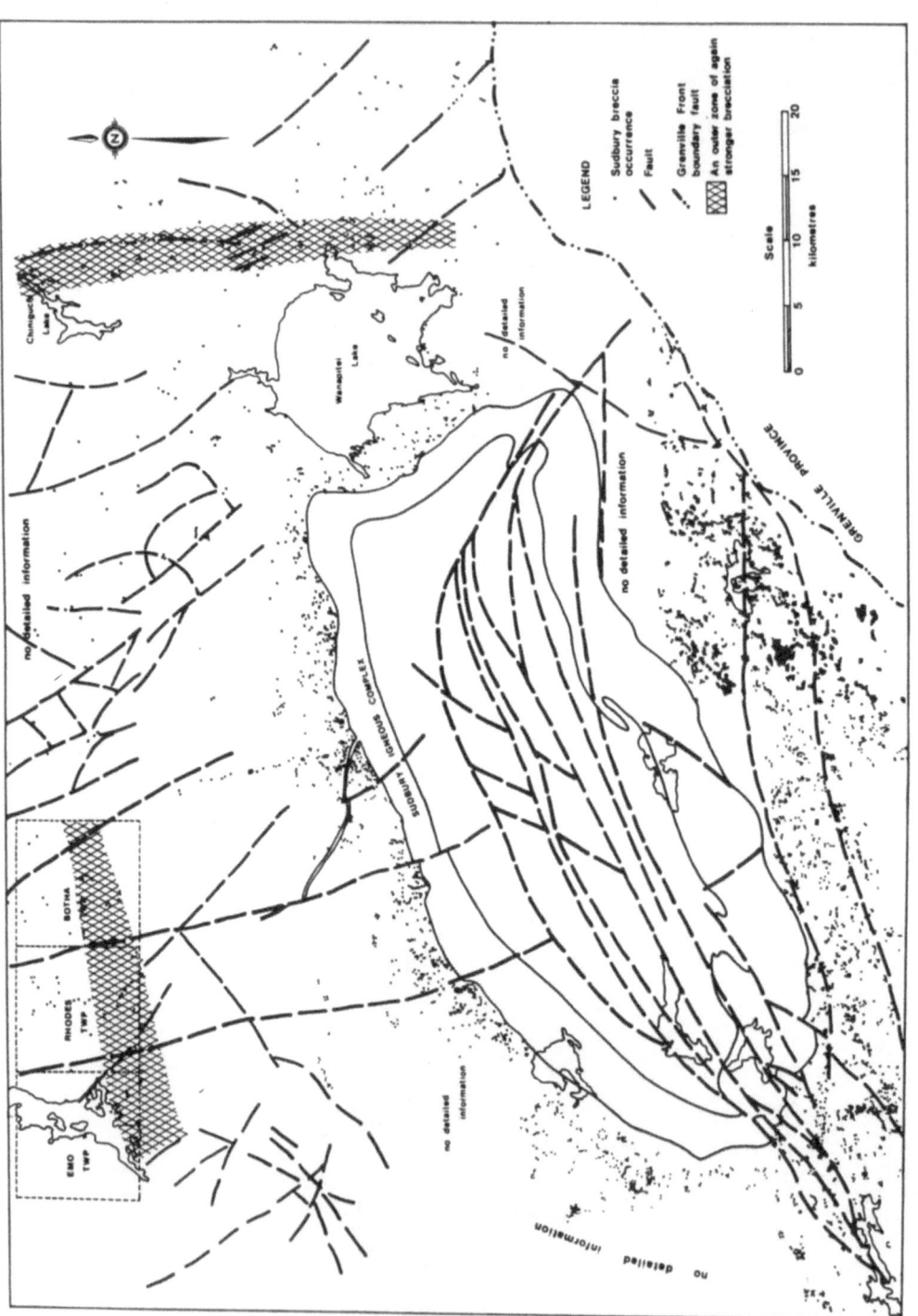

Fig. 6
Distribution of Sudbury Breccia bodies around the Sudbury Igneous Complex (after Dressler, 1984).

rous, irregular (Fig. 5), or tabular bodies around the entire Sudbury Igneous Complex and is abundant up to 5 to 10 km from the lower contact of the Complex but is known to occur much farther away. Evidence exists for the presence of 2 specific zones of increased brecciation about 20 to 25 km and 80 km away from the contact of the Complex. In Fig. 6 only the first outer zone of strong brecciation is shown. In this figure, breccia bodies (large and small) appear to be more common south of the Complex than to the north. This apparent difference may merely be a function of rock exposure which is greater in the south than in the north.

Breccia bodies commonly follow structural weaknesses, such as rock contacts, foliations, joints and faults. The largest body of intense brecciation is approximately 11 km long, up to 0.5 km wide, and, in general, follows the trend of Huronian lithologies south of the Igneous Complex. The orientations of many irregularly shaped breccia bodies is not known. Most breccia dikes, in both the North and South Range footwall, dip steeply or vertically. Over 700 measurements of breccia dikes (Dressler, 1984) do not show any apparent correlation of their orientation with respect to the shape of the Sudbury Structure.

Breccia contacts with the host rocks are sharp. Most fragments in the breccias are rounded or subrounded, the very small rock and mineral fragments, however, are subangular to angular. The size of the fragments depends on the size of the breccia bodies. The fragments, in general, are of the same rock type as the host rock. Other fragments, however, are common as well and may be derived from rocks nearby, or sources not exposed on the surface.

The breccia matrix is commonly gray or black and consists of a microscopic and submicroscopic, weakly recrystallized, massive rock flour. Vesicular and non vesicular, recrystallized melt matrices, igneous-textured and cataclastic flow banded matrices also occur but are less common than the massive variety. The Sudbury Breccia grades into Footwall Breccia (Fig. 7) over a distance of about 5 to 20 m.

Speers (1957) stated that "the chemical composition of the breccia matrices reflects the composition of the host rocks and also indicates that foreign material has been mixed with material from the host rocks". Dressler (1984) chemically compares the breccia matrix with host rocks and host rock fragments from many locations around the Igneous Complex. His results, some of which are shown in Fig. 8 and 9, indicate clearly that the chemical characteristics of the breccia matrices can be explained by assuming that the breccia forming process was not an entirely in situ process, but that some transport of

Fig. 7
Transition of Sudbury Breccia into Footwall Breccia. Notice dissolution of Sudbury Breccia vein (E-W trending in centre of photograph; diameter of lens cap is 5 cm). Northern footwall of Sudbury Igneous Complex.

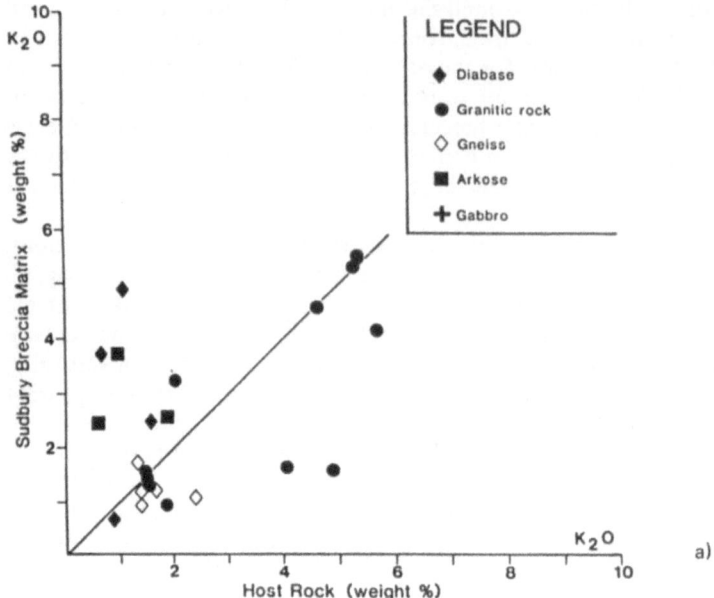

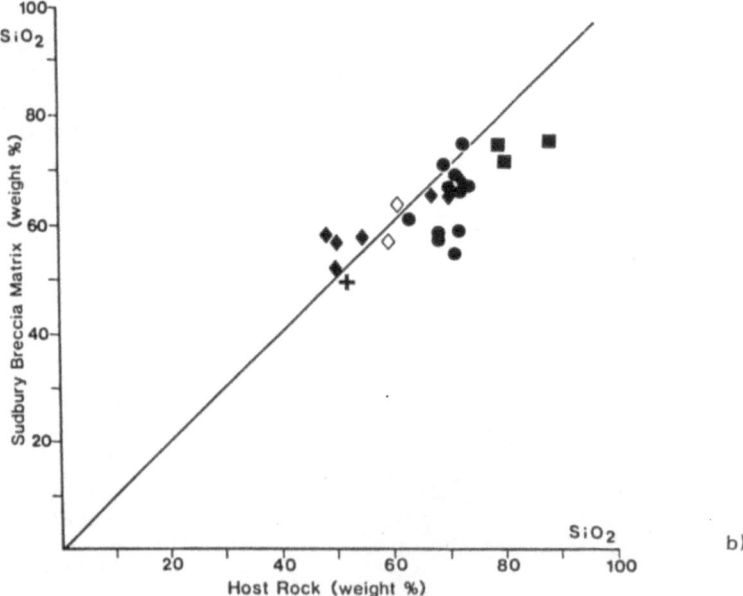

Fig. 8

Sudbury Breccia matrix vs. host rock. NB.: There is no systematic
increase or decrease in K_2O content of the breccia matrix. Analyzed
samples from all around the Sudbury Igneous Complex
(after Dressler, 1984).

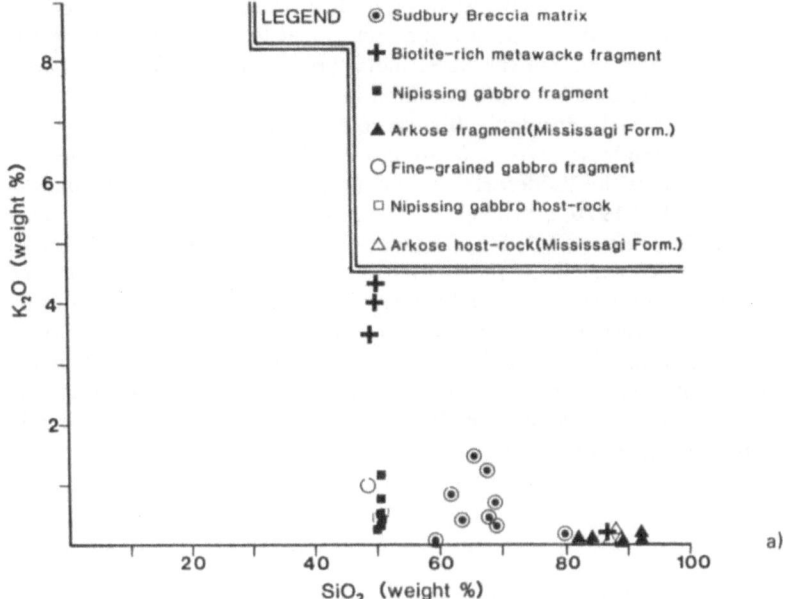

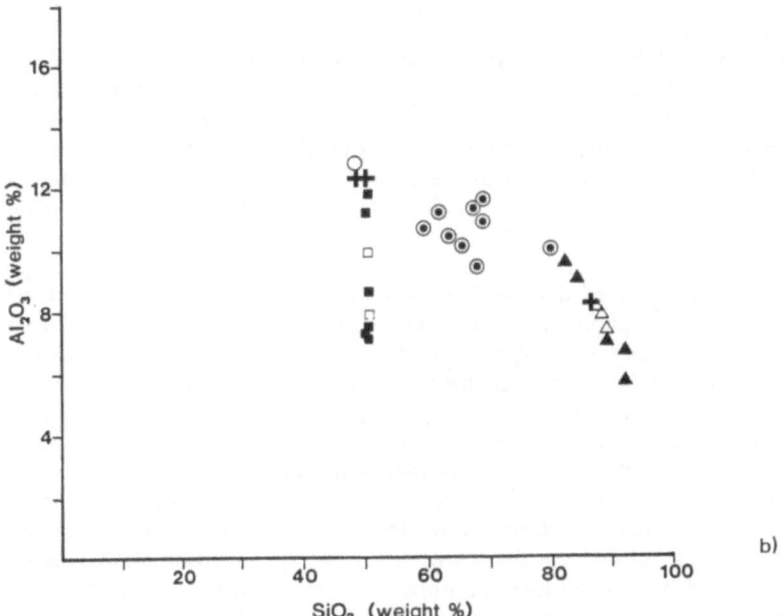

Fig. 9

Plots of SiO_2 vs. K_2O and Al_2O_3. Sudbury Breccia, near Laurentian
University Campus, Sudbury. Breccia matrix plots between hostrocks
and fragments. Plots of other chemical data are similar. There is no
addition of chemical components from outside the breccia body
(after Dressler, 1984).

pulverized material occurred. Breccia matrices in mafic host rocks such as diabase indicate incorporation of some granitic or arkosic material, or at least, rocks closer to the felsic end of the chemical spectrum. On the other hand, breccia matrices in arkosic host rocks are enriched in elements that are more common in more mafic rocks, and thereby suggest incorporation of such rocks during the breccia-forming process.

The Sudbury Breccias are pseudotachylites and are comparable with pseudotachylites in other impact structures such as the Vredefort Structure (Wilshire, 1971), the Manicouagan Structure in Quebec (Dressler, 1970, Murtaugh, 1976) and also with breccias in the basement rocks of the Ries (Stöffler et al., 1977). Their origin is related to the formation of the Sudbury Structure as deduced from the spatial relationship of the breccias around the Sudbury Structure. The breccias originated mainly by brittle fracture, comminution and forceful injection of pulverized material into the footwall rocks in an explosive event related to the formation of the Sudbury Structure and later readjustments of the crust. Frictional heat locally led to partial or complete melting of crushed rock material. Shock metamorphic features in the breccias relate the breccia forming processes directly to processes caused by the impact of a meteorite.

Footwall Breccia

The Footwall Breccia contains much of the sulphide mineralization of the Sudbury mining area especially on the North Range, and as "leucocratic breccia" formerly had been included in the Sublayer (Pattison, 1979). The breccia forms discontinuous lenses and sheets along the lower contact of the Sudbury Igneous Complex. It is most common in the East and North Ranges.

The contacts of the Footwall Breccia with the norite or the Sublayer of the Igneous Complex are abrupt. Pattison's (1979) "megabreccia" zone underlies the Footwall Breccia and is characterized by autochthonous footwall rock fragments up to several 10s of metres in size, embedded in, and intruded by the Footwall Breccia. The breccia forms small offshoots up to 250 m away from the main Footwall Breccia bodies. It also occurs in the Sudbury Offsets where it was probably passively transported by the Sublayer intrusion (Fig. 10).

The Footwall Breccia (Fig. 11) is a heterolithic breccia characterized by a variety of angular to subrounded fragments of diverse sizes. The fragments are derived from local terrane and are embedded in a pinkish gray, gray or dark gray matrix described by Pattison (1979) as "mosaic granoblastic metamorphic". This characterization, however, does not encompass all microscopic features of the breccia, such as relict shock metamorphic features, decussate stubby plagioclase recrystallization, granophyric intergrowth of quartz and sodium and potassium feldspar and porphyroblasts of quartz, plagioclase, potassic feldspar, biotite, hornblende, orthopyroxene, clinopyroxene and magnetite.

The Footwall Breccia components have been affected by several geological processes. The brecciation and shock metamorphism are related to the meteorite impact. Contact metamorphic overprint was caused by the intrusion of the Igneous Complex. This together with a later period of regional metamorphism, obliterated many primary features. Granitic and granophyric mobilizates invade the breccia bodies, as small irregular patches, stringers and dikes.

The authors believe that the Footwall Breccia originally consisted of a predominantly parautochthonous mass of crushed, and pulverized, in part, shock metamorphosed rocks. In many aspects the Footwall Breccia was originally very similar to Sudbury Breccia, into which it appears to grade (Fig. 7). This apparent gradation, however, is not an indication

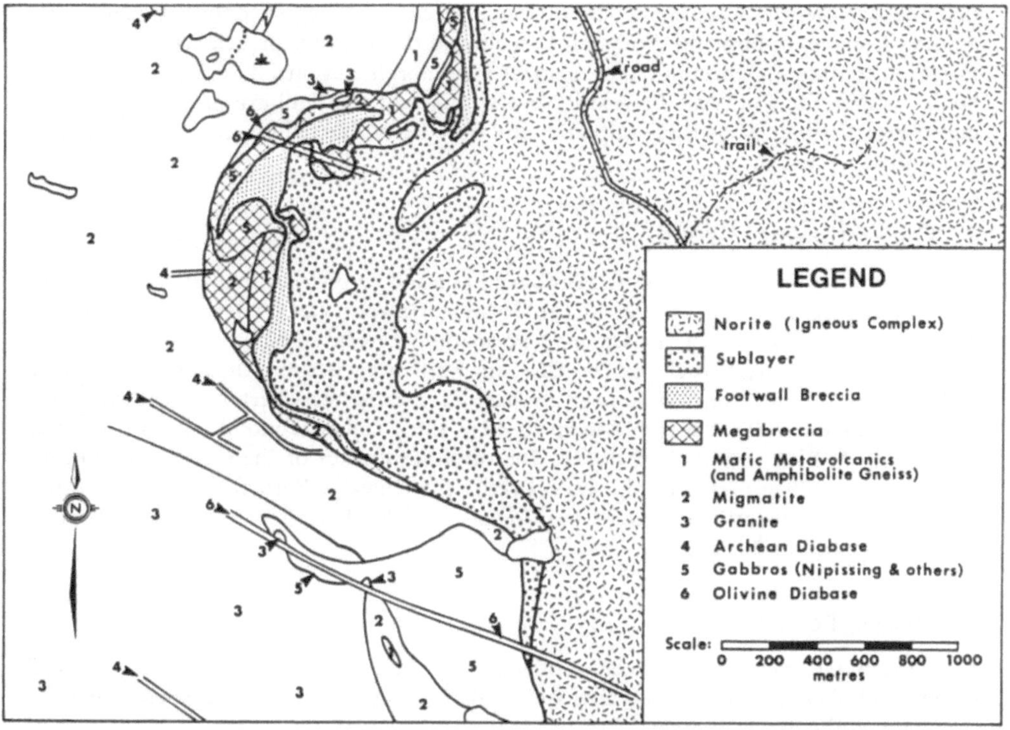

Fig. 10
Field relationships of Footwall Breccia, megabreccia and Sublayer in the Trillabelle Embayment of the North Range of the Sudbury Structure.

Fig. 11
Footwall Breccia. At Strathcona Mine, Levack Township, north of Sudbury Igneous Complex.

that the two breccias were formed by the same process but suggests only that close to the Igneous Complexes both the Footwall Breccia and the Sudbury Breccia were affected by the same contact metamorphic and regional metamorphic events. The large sulphide ore bodies within the Footwall Breccia in several mines of the Sudbury district were introduced into the breccia as fluid sulphides and have been affected by tectonism simultaneous with this process.

3. The Whitewater Group

The Whitewater Group occurs in the Sudbury Basin and consists of three lithostratigraphic units which, from bottom to top, are the Onaping, Onwatin and Chelmsford Formations. The heterolithic breccias of the Onaping Formation are central to any discussion on the origin of the Sudbury Structure. The Onwatin Formation is approximately 600 m thick and consists of mudstones and near its base of a carbonate-chert unit known as the Vermilion Member. This member is up to 45 m thick and is economically important as it contains Zn-Pb-Cu sulphide deposits. The turbidite wackes of the Chelmsford Formation of about 900 m preserved thickness overlie the Onwatin Formation. These two formations are not directly related to the origin of the Sudbury Structure and, therefore, are not discussed in this paper.

The Onaping Formation

The Onaping Formation has been subdivided by Peredery (1972a) and by Muir and Peredery (1984) into four units which are the Basal Member, the "Melt Bodies", the Gray Member and the Black Member.

Basal Member

The Basal Member (Fig. 12), at the base of the formation, is discontinuous, up to 200 m thick, and intruded by the Sudbury Igneous Complex. On the South Range it consists essentially of fragments of Huronian quartzofeldspathic metasediments with very minor granitic and rare anorthositic fragments, also derived from South Range footwall rocks. On the North and East Ranges the Basal Member is comprised of a mixture of various proportions of metasedimentary, granitic, gneissic and relatively rare mafic rocks. These rocks are similar to rocks found in the adjacent terranes.

The size of the fragments ranges from a few millimetres to several hundred metres in diameter. In any given area the fragments may be very well sorted, and on the average are less than one metre in diameter. The contact between the Basal Member and the overlying Gray Member is either sharp or gradational over generally less than one metre. Contacts of the Melt Bodies with the granophyres of the Sudbury Igneous Complex and with the Grey Member are sharp, those with the Basal Member are sharp in one place, gradational in another.

The matrix of the breccias of the Basal Member is strongly recrystallized and inhomogeneous. Originally it probably consisted of finely pulverized material. Stevenson (1963) recognized the pulverized character of the matrix in the "quartzite breccia" of the South Range and described it as in part permeated by a granophyre-rich fraction. No recrystallized glass fragments typical of the Gray and Black Members have been observed in the Basal Member.

French (1968), Peredery (1972b) and Muir and Peredery (1984) recognized shock metamorphic features in mineral and rock fragments.

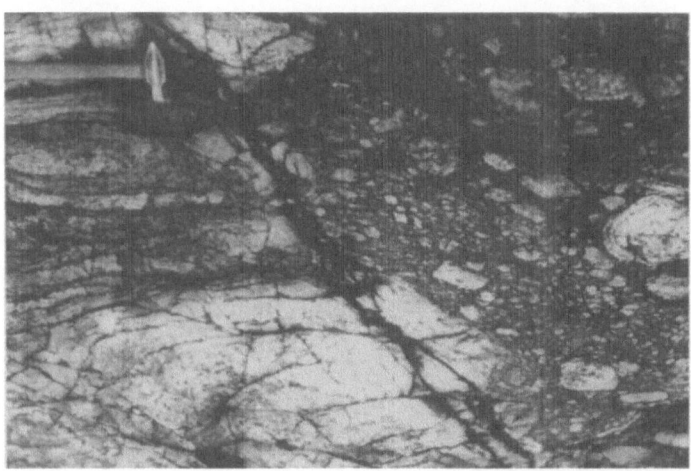

Fig. 12
Basal Member of Onaping Formation. Almost all fragments are meta-quartz arenite. Left half of photograph shows part of a large block with weakly deformed bedding. Hammerhead is 13 cm long. (Photograph courtesy of T. L. Muir).

Gray and Black Members: General Description and Stratigraphy

The Gray Member overlies the Basal Member and in turn is overlain by the Black Member. Both upper members consist of a heterogeneous mixture of various recrystallized glasses and country rock fragments embedded in a very fine-grained matrix of recrystallized glass fragments, pulverized country rocks, mineral fragments and submicroscopic material. The Gray Member is characterized by a lack of stratification in most of the deposit (discontinuous stratification and contacts between breccia units in the upper member have been described by Muir and Peredery, 1984), the Black Member by the presence of carbonaceous material in its matrix. In both members (Fig. 13 and 14) recrystallized glasses are fragmented and of heterogeneous nature. Country rock fragments in both members commonly have a variety of shock metamorphic features.

The Gray Member is about 700 m thick. Well defined bedding is absent from most of the member and some indication of stratification is indicated by discontinuous horizons of recrystallized and boulder size country rock fragments (Peredery, 1972a). Close to the boundary with the overlying Black Member, the Gray Member grades into a layer rich in shards and poor in rock fragments. This layer is continuous and up to several tens of metres thick.

The Black Member is characterized by a dark gray to black carbonaceous matrix. Discontinuous bedding and discontinuous boulder size rock and recrystallized glass fragment trends are present in the member (Peredery, 1972a, b; Muir, 1982; Muir and Peredery, 1984). The lower half of the member, in general, contains a chaotic assemblage of recrystallized glass and rock fragments similar to that of the Gray Member. In the upper half, average fragment size decreases and the unit becomes progressively finer grained. The contact with the overlying mudstones of the Onwatin Formation is gradational.

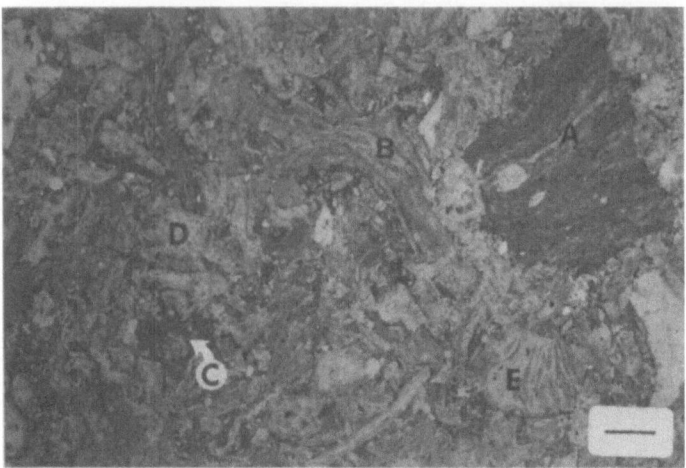

Fig. 13
Gray Member of Onaping Formation. Fragment-supported heterolithic breccia.
A – Fluidal fragment
B – Complex felsic fragment

C – Complex mafic fragment
D – Simple felsic fragment
E – Banded, simple felsic fragment

Plane polarized light. Scale: 2 mm
(Photograph courtesy of T. L. Muir)

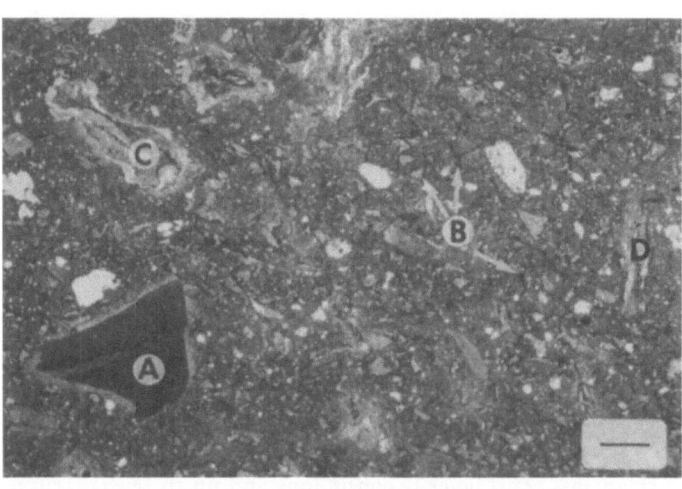

Fig. 14
Black Member of Onaping Formation from about 1 m above contact with Gray Member.
The matrix is grey-green and contains few carbon particles.
A – Black, siltstone (?) fragment
B – Shard-like fragment composed of chlorite
C – Complex mafic fragment composed of chlorite with feldspathic, discontinuous rim.
D – Felsic complex, recrystallized glass.
Plane polarized light. Scale 2 mm
(Photograph courtesy of T. L. Muir)

Recrystallized Glasses

Two major types of glasses occur in the Onaping Formation, the "complex glasses" and the "fluidal glasses" (Peredery, 1972a, b Muir and Peredery, 1984). All glasses are recrystallized.

The complex glasses are macroscopically heterogeneous and commonly exhibit a typical grayish to bluish colour. They have subequant shapes and are up to several centimetres in size. Minor plastic deformation is indicated by the shapes of some complex glass fragments. The glasses consist of devitrified domains up to several millimetres in diameter of quartz or feldspar or less commonly devitrified mafic minerals that chemically correspond to calcic clinopyroxene (Peredery, 1972a, b). Diaplectic planar features in relic felsic minerals are present in some complex glass fragments. This observation and the heterogeneous character of the glasses suggests that the complex glasses represent severely shocked (maskelynite state), recrystallized country rock fragments. Plastic deformation or partial "normal" melting in some complex glasses represent a transition from "complex glasses" to "fluidal glasses".

These fluidal glasses are inhomogeneous and commonly contain rock and mineral fragments. They are composed of finely recrystallized material with fluidal lines or domains that consist of variable proportions of felsic and mafic products. Electron microprobe analyses yield a wide range of compositions in any given sample and support microscopic observations (Peredery, 1972a, b). The fluidal glasses are interpreted here to represent normal shock melted rocks that flowed but did not become completely homogenized, i. e. glasses in which the original mineral lattices completely collapsed due to the shock induced heat that affected the rocks in the impact target area.

Shards completely replaced by chlorite and titanite, or recrystallized to quartz, feldspar, and amphibole and chlorite, may represent complex or fluidal glasses that originated from shock melted intermediate to mafic, originally fine-grained target rocks that were fragmented and airborne in the impact process.

Welding of glass fragments, especially of very fine material, is commonly absent in the breccias of the Onaping Formation. However, some welding between large fragments of country rocks and glasses in bomb-shaped bodies has been observed. Glass to glass welding occurs in some large pancake or irregularly shaped bodies.

Country Rock Fragments and Shock Metamorphism

Rock fragments in the Onaping Formation range in size from a few millimetres to several hundred metres in their largest dimension, are angular to subrounded in shape and make up at least 10 percent, in places 30 percent of the Gray and Black Members of the formation. They consist of arkoses, quartz arenites, wackes, and conglomerates derived from the Huronian Supergroup and of Archean granites, gneisses and anorthositic rocks. Amphibolite and gabbro fragments are present but are not common. Peredery (1972b) described a breccia fragment in the Onaping Formation, similar in appearance to Sudbury Breccia.

Shock metamorphic features observed in country rock fragments include devitrified diaplectic glasses of feldspar and quartz, planar features in quartz and plagioclase, plastic deformation in quartz and feldspar and partial melting of biotite, titanite and zircon (French, 1968; Peredery, 1972b). Some rock fragments exhibit shatter cones.

Melt Bodies

Melt rocks occur as irregular shaped bodies close to the contact of the Basal Member with the Gray Member. They occur also as lenses or dikes in the Gray Member and less commonly as lenses or dikes in the Black Member. They were interpreted by early investigators as lava flows, volcanic breccias and agglomerates (Burrows and Rickaby, 1929; Moore, 1930; Cooke, 1946) or as a chilled phase of the Sudbury Igneous Complex (Stevenson, 1963; Muir, 1984).

The rocks are interpreted as impact melts and field evidence suggests that they behaved as intrusions in one place, or as extrusions in another (Peredery and Morrison, 1984). The bodies exhibit chilled or chilled-brecciated margins and are fine grained, igneous-textured. They contain abundant country rock xenoliths such as granite, gneiss, metasediment and, less commonly, mafic rocks. The angular to rounded fragments range in size from a few millimetres to a few tens of centimetres and exhibit shock metamorphic features. In any given body, they are generally well sorted as to size.

The close field relationship between the melt rocks and the Basal Member and the similarity between the fragments in the Basal Member and in the Melt Bodies suggests that the Basal Member was the source of the inclusions in the melt rocks, or that the fragments in both the Basal Member and the melt rocks were directly derived from the same footwall terrane.

Chemical composition of the melt bodies ranges from rhyolite to andesite and to lime poor rocks chemically similar to keratophyre (Burrows and Rickaby, 1929; Thomson, 1956; Cooke, 1946). Peredery (1972a, b) confirmed the earlier chemical investigations and showed that the melt rocks may range from rhyolitic to andesitic compositions within the same body.

In a recent study Krogh et al. (1984) demonstrate striking morphological and isotopical similarities between shocked zircons from Archean basement gneisses from just north of the Sudbury Igneous Complex and those in a granitic clast and in samples from fluidal glasses and from a melt rock of the Onaping Formation. These observations, in Krogh et al.'s and also our view support the impact origin for the melt rocks and other lithologies of the Onaping Formation.

4. The Sudbury Igneous Complex

General Characteristics

The Sudbury Igneous Complex has an elliptical outline (Fig. 1), is approximately 60 km long and 25 km wide, and consists of the Lower, Middle and Upper Zones of the Main Mass and of the Sublayer. The Complex has been dated at 1.850 Ma (Krogh et al., 1984). The Lower Zone consist of mafic norite, quartz-rich norite, "South Range norite" and felsic norite. It is overlain by the quartz gabbro of the Middle Zone and the granophyres of the Upper Zone. Not all of these units are continuous. The mafic norite occurs only in the North Range where the felsic norite makes up the bulk of the Lower Zone. The quartz-rich and South Range norites are restricted to the South Range. Quartz gabbro and granophyres occur throughout the Complex (Fig. 15).

Along the base of the Lower Zone there are discontinuous bodies of Sublayer consisting of distinctive mafic igneous rocks that are characterized by sulphide mineralization and mafic, ultramafic and footwall rock inclusions. The Sublayer forms offshoots, known as "offsets", that intruded the footwall rocks and, in a few places, also the lower section of the norite of the Main Mass.

The Sudbury Igneous Complex is similar to other large igneous complexes such as the Stillwater, and Bushveld. There are, however, major differences that make the Sudbury Igneous Complex unique. It differs from nearly all other layered complexes in two important aspects (Naldrett and Hewins, 1984): 1) There is no graded or fine scale igneous layering; 2) All rocks exposed at the surface and in mine workings are very rich in SiO_2. The high SiO_2 content is not attributable to fractional crystallization alone and Irvine (1975) and Naldrett and MacDonald (1980) proposed that the magma of the Igneous Complex had become contaminated by SiO_2-rich country rocks.

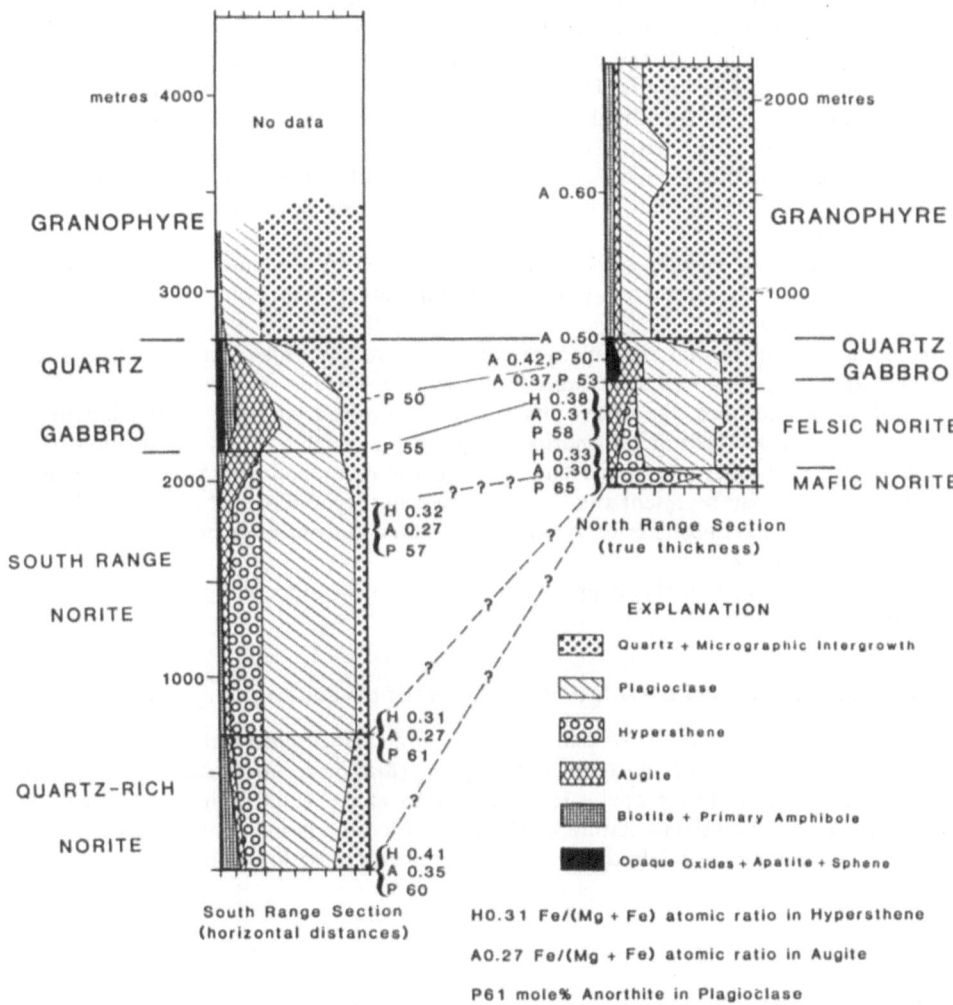

Fig. 15

Mineralogical variations on sections through the North and South Ranges of the Sudbury Igneous Complex (after Naldrett et al., 1970).

A voluminous literature exists on the Main Mass of the Sudbury Igneous Complex. More recent studies are those of Naldrett and Kullerud (1967), Naldrett et al. (1970, 1972), Gasparrini and Naldrett (1972), Peredery and Naldrett (1975), Gibbins and McNutt (1975) and Kuo and Crocket (1979).

Naldrett and Hewins (1984) summarize much of the published literature and state that three main models have been proposed for the Igneous Complex; 1) the Complex is a folded differentiated sill; 2) both the norite and granophyre intruded separately as ring dikes and; 3) the Complex is a funnel shaped intrusion. Naldrett and Hewins (1984) further state that the Complex is not a ring dike complex as it has been document- ed that cryptic variation of pyroxene and plagioclase compositions extend across the Lower and Middle Zones. In the North Range phase layering parallel to the footwall/ Igneous Complex contact suggests that the body is a sill. However, as Naldrett and Hewins (1984) further point out, "the irregular distribution of plagioclase-rich layers in the granophyre, coupled with irregularities in the upward variation in the Fe/(Fe + Mg) ratio of augite through the granophyre, indicate that the complex is not a simple in situ differentiate. The very felsic average composition, if this is calculated on the basis of the geometry of a folded sill, also argues against such a simple interpretation. The ob- servation that igneous lamination on the South Range dips north at a shallower angle than the footwall of the complex should be treated with caution as an indicator of the attitude of phase layering, but is suggestive that this part of the intrusion resembles a funnel rather than a sill."

Souch et al. (1969), Hewins (1971) and Pattison (1979) describe the Sublayer and Naldrett et al. (1984) summarize published results and present some new observations made on the Sublayer.

There is some evidence that the Sublayer consists of at least three intrusive phases, two of which appear to be post-Lower Zone in age as they intrude the norite. No clear field relationships of the third, the oldest phase with the norite have been observed. This fa- cies, therefore, may be pre-or post-norite in age. It is characterized by mafic, ultramafic and minor footwall rock inclusions. The origin of the mafic and ultramafic inclusions is not known. They may represent the lowest, buried and layered part of the Igneous Com- plex itself or may have been derived from a deep seated igneous body not directly related to the Igneous Complex.

The Sublayer is the main nickel-copper sulphide ore bearing unit of the Sudbury Structure (Neldrett, 1984).

Chemical Composition and the Origin of the Magma of the Sudbury Igneous Complex

Fig. 16 summarizes the results of published and unpublished chemical investigations on the norite and Sublayer of the Igneous Complex. (For more information see Naldrett and Hewins, 1984; Naldrett et al., 1984, and Rao et al., 1983). These investigations may shed light on the original nature of the magmatic liquids and may also prove or dis- prove that the Complex is solely the product of an in situ crystallization process.

Fig. 16

Chemical characteristics of the Main Mass norites and the Sublayer.
In the MgO vs. oxide and compatible element (Sc, Cr, Co) diagrams the solid circles represent the Sub- layer samples from the South Range and the open circles represent the North Range samples. "r" is the regression coefficient. In the other diagrams Main Mass norites and Sublayer data are combined. ▶

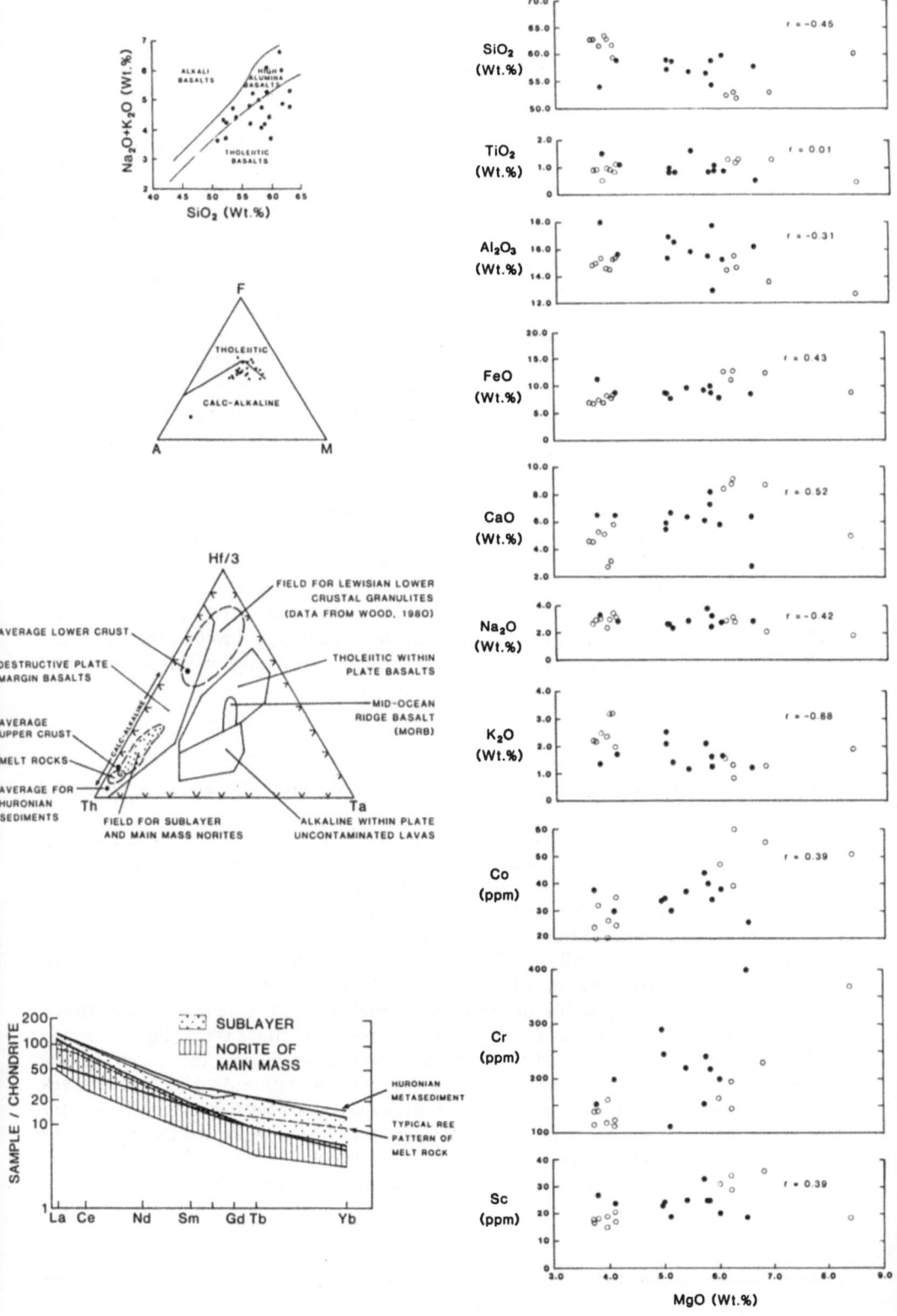

The norites of the Lower Zone and the Sublayer are silica saturated or silica oversaturated. Their normative quartz content ranges from about 1 to 20 percent and the normative hypersthene content ranges from 14 to 30 percent. The norites are characterized by a K_2O content that is higher than that of continental tholeiites with similar Mg number. The $Na_2O + K_2O$ vs. SiO_2 plot demonstrates the general tholeiitic nature of the rocks. On an AFM diagram the compositions plot in the field of calc-alkaline rocks. There is a poor correlation of MgO with other oxides and compatible elements (Cr, Co, Sc). This indicates that the rocks did not result from the crystallization of a single fractionating liquid. Further evidence against simple fractional crystallization is based on magmatic discrimination diagrams like the Th-Hf-Ta diagram (Wood, 1980). In this diagram the rocks of the Igneous Complex plot in the field of calc-alkaline rocks and are believed to have been formed by extensive assimilation of crustal material.

The norites and Sublayer rocks show similar REE patterns and are characterized by high absolute REE contents compared to many continental tholeiites. Their chondrite-normalized patterns show strong LREE enrichments and no HREE fractionation. It is significant to note that the total REE content in the mafic rocks of the Igneous Complex is very high for cumulus mafic rocks. If one models the liquid interstitial to the cumulus pyroxene and plagioclase one has to assume that the liquid was extraordinarily enriched in LREE. It, therefore, was probably strongly contaminated with crustal material. Gibbins and McNutt (1975) compared the $^{87}Sr/^{86}Sr$ ratios of the Nipissing gabbro with those of the Igneous Complex and concluded that both rock units probably originated from the same or a similar source. Recent geophysical investigations (Gupta et al., 1984) indicate the existence of a large mafic body at depth below the Igneous Complex. This body probably acted as a source magma chamber for the Nipissing gabbroic intrusions in the Sudbury area. The meteorite impact probably initiated remelting of this source body which resulted in the generation of the Sudbury magma.

5. Impact Model

Our impact model is based mainly on Peredery and Morrison (1984) and is illustrated in Fig. 17.

At the time of impact (Fig. 17a) the Sudbury area was covered by volcanic and sedimentary rocks of the Huronian Supergroup and possibly was under water. No accurate estimate of the size (possibly 2—3 km in diameter), density (possibly 3—3,5 t/m^3), and the velocity (possibly 15—25 km/s) is possible. The intense shock due to the impact caused melting and vaporization of the target rocks and the bolide and brecciation (Sudbury Breccia and Footwall Breccia) and shock metamorphism of the Footwall rocks. Elementary carbon found in the Black Member of the Onaping Formation and in the Onwatin Formation was possibly produced in a shock-induced reduction process affecting the air masses at the target area[2]. During the transient crater stage (Fig. b—c) melt rock and shocked and unshocked rock fragments were ejected. Near the target area the explosion caused a quasi-vacuum (stage between Fig. 17c and d) into which first fall-back debris was deposited. Air and possibly water masses carrying rock debris returning to the target at a speed of several hundred km/hour eroded already deposited material in one place and deposited rock and glass fragments in another (Fig. 17d). Contacts observed in the Onaping Formation may have their origin in these processes. Mixing of rock and glass

---

2 Shock-induced reduction ($SiO_2 \rightarrow Si$) has experimentally been documented by Jammes et al. (1983).

Fig. 17

The development of the Sudbury Structure (see text; Sudbury Breccia is shown only in Figure c); from Peredery and Morrison (1984).

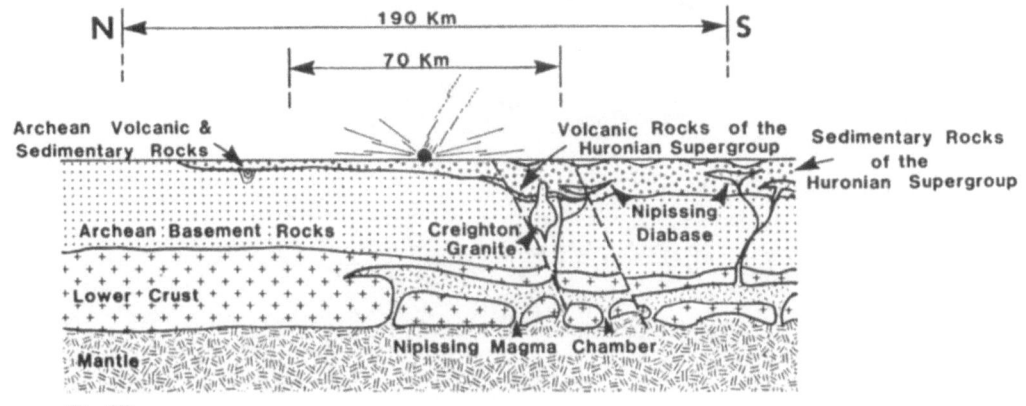

Fig. 17a

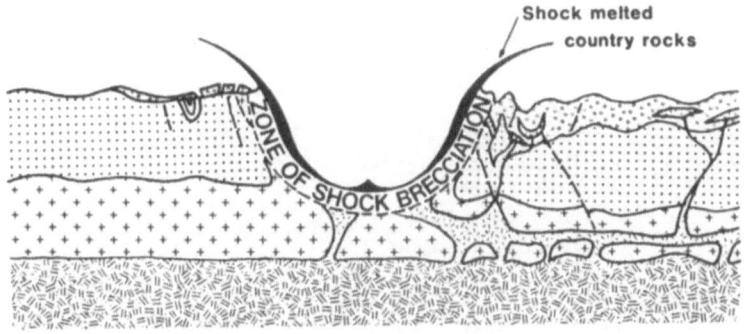

Fig. 17b

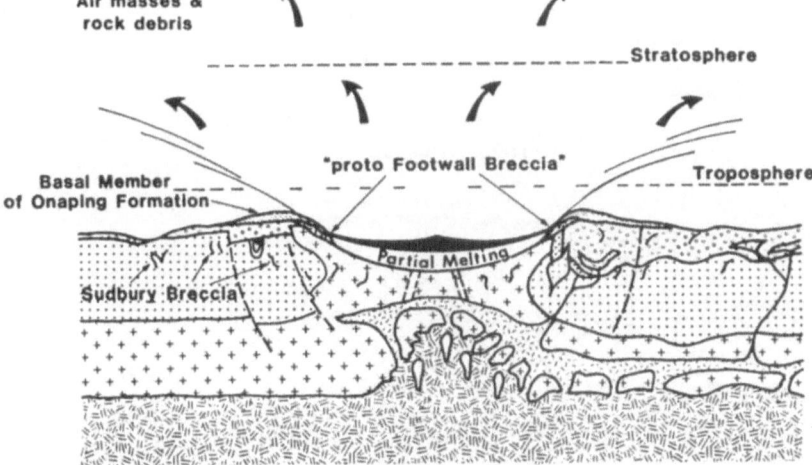

Fig. 17c

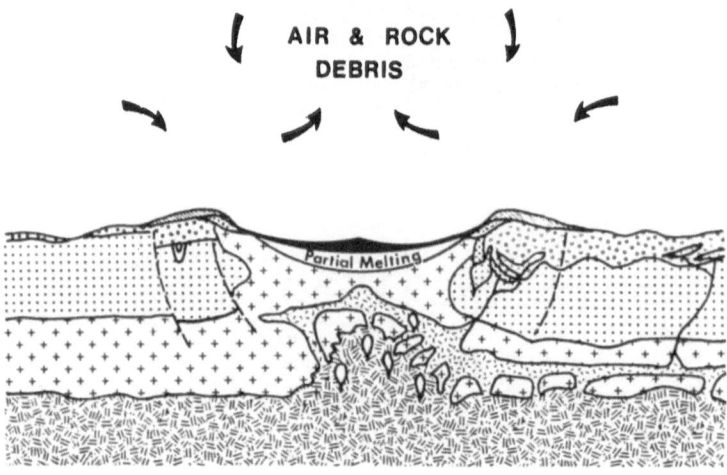

Fig. 17d

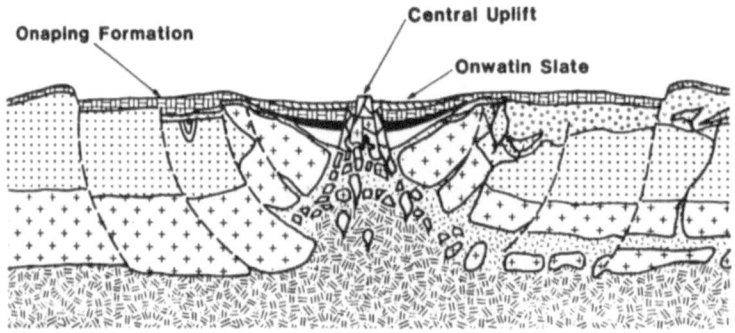

Fig. 17e

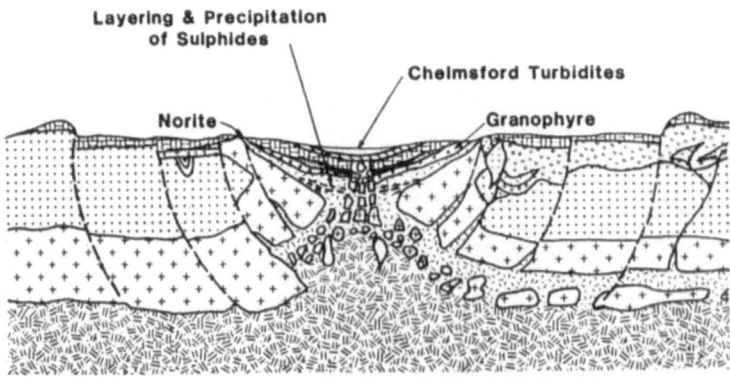

Fig. 17f

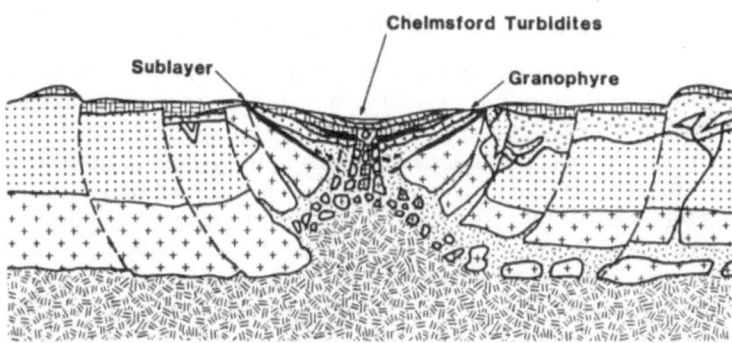

Fig. 17g

fragments during ballistic transport and deposition and redeposition processes produced multiple breccias essentially in one process.

The Basal Member of the Onaping Formation represents the earliest fallback and possibly also brecciated country rock that ground-surged along the crater wall. The Gray Member is interpreted as the main fallback breccia and the accumulation of fine-grained, shard-like debris in the upper part of this member as material that took longer to fall back and therefore became fairly well sorted. Air and water masses (Peredery and Morrison's (1984) tsunami wave), including elementary carbon, carried material from outside the crater and deposited it on the Gray Member, thereby producing the Black Member. The lithostatic load resulting from the fallback forced the underlying impact melt up along the crater wall to form irregular lenses and dikes in the fallback breccias.

Reworking of the Onaping Formation produced the mudstones of the Onwatin Formation.

In response to the intense and major adjustments, the area peripheral to the crater collapsed to produce the Sudbury crater basin measuring about 190 km in diameter. A central uplift (Fig. 17e) was possibly formed in the crater in response to the elastic floor rebound and subsequent adjustments. This crater collapse and readjustments of the crust produced further breccias in the footwall.

Fracturing of the crust, excavation of the crater and isostatic rebound and adjustments resulted in basic magma rising from depth, possibly from a reactivated Nipissing magma chamber, and intruding the crater thereby producing the Sudbury Igneous Complex. SiO_2-contamination of this magma caused sulphides to "drop-out" and accumulate in the lower portions (Fig. 17f) of the Igneous Complex from where it became emplaced along the crater wall and in the "offsets" during the final magmatic (Sublayer) phases of the Igneous Complex (Fig. 17g). Collapse of the magma chamber caused the crater walls to steepen and in the same process the central uplift collapsed. The wackes of the Chelmsford Formation were deposited prior and after first faulting and deformation of the Sudbury Structure.

Comparison of the Sudbury Structure with the Ries Crater, Germany

The Ries is a well preserved and intensively studied impact crater in southern Germany. In Fig. 18 the crater stratigraphy of the Ries is compared with that of the Sudbury Structure. Several similarities are obvious.

The basement of both craters is strongly brecciated. Anastomosing breccia dikes very much like the Sudbury Breccia pseudotachylites also occur in the Ries (Bayer. Geol. Landesamt, 1974, Dressler et al., 1969). The base of the breccia sequences (Footwall Breccia and Basal Member of the Onaping Formation in Sudbury and the Bunte Breccia of the Ries) consist of relatively weakly shocked rock fragments and a large amount of fragments from the original sedimentary cover rocks suggesting a reversal of the original stratigraphy of the target area in the breccia sequences at both craters. These basal units are overlain by the suevite in the Ries and the Gray and Black Members of the Onaping Formation in Sudbury. All these units consist of diaplectic, heterolithic and glass rich breccias, that contain lesser amounts of fragments from the original sedimentary cover of the target area than the basal breccia units.

The authors are aware that the Bunte Breccia is absent from the central crater cavity in the Ries, that it represents a ballistically transported breccia unit, and that the crater suevite possibly is not a fall back breccia. In the Sudbury Structure a central crater cavity is nowhere exposed. Nevertheless the authors are impressed by the similarity of the breccia stratigraphy of the Ries and Sudbury Structures.

No igneous intrusion is associated with the Ries as the less catastrophic Ries impact did not trigger any observable magmatic activity. The rocks in Sudbury have all been affected by subgreenschist to greenschist facies regional metamorphism and near the Sudbury Igneous Complex by contact metamorphism. In the Ries the breccias have been affected by diagenetic processes only.

6. Evidence Forwarded Against an Impact Origin for the Sudbury Structure

Several authors have pointed out that the Sudbury Structure lies in an unique position near the present "junction" of three structural provinces of the Canadian Shield, namely the Superior, Southern and Grenville Provinces, and that it is located at the junction of regional fault systems (Card and Hutchinson, 1972; Muir, 1984). Another observation, interpreted as evidence against an impact origin for the Sudbury Structure, are paleocurrent measurements in rocks of the Chelmsford Formation (Cantin and Walker, 1972) which would indicate that the Sudbury Basin never was circular, but elongate with a regional west-southwesterly dip during deposition of the wackes of the Chelmsford Formation.

The Sudbury Structure is located on a large gravity high and on a "linear arrangement of positive aeromagnetic anomalies". This situation is believed by Muir (1984) to be too significant to be the location of a fortuitous meteorite impact. The Wanapitei Lake Structure located at Wanapitei Lake just east of the Sudbury Igneous Complex (Fig. 1) and interpreted by Dence and Popelar (1972) as another impact site, conforms to the shape of the East Range Sudbury Igneous Complex. This would be too coincidental a configuration to be accounted for by meteorite impact (Muir, 1984).

The observations listed above and forwarded as proof of an endogenic origin for the Sudbury Structure have been discussed by Peredery and Morrison (1984). For a detailed account, including a discussion of observations both in favour and against an impact origin for Sudbury the reader is referred to the Ontario Geological Survey publication "The Geology and Ore Deposits of the Sudbury Structure" (Pye, Naldrett and Giblin, 1984).

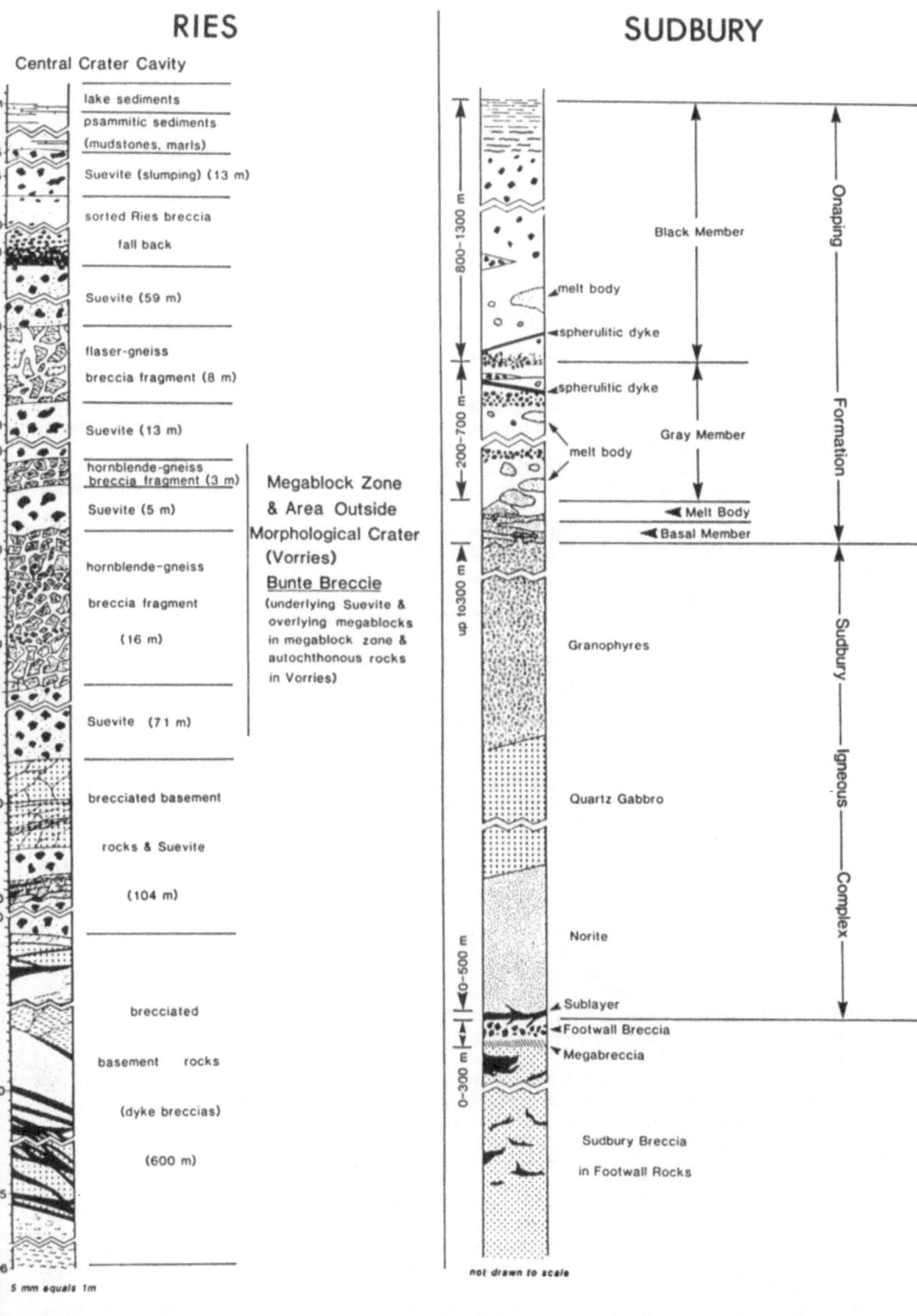

Fig. 18

The Sudbary Structure compared with the Ries Crater, Germany.
(Ries Crater: Bayer. Geol. Landesamt, 1974; Sudbury Structure, Onaping Formation: Muir and Peredery, 1984).

Acknowledgments

The authors would like to thank G. Johns and J. Wood for their careful and critical reading of the manuscript. They would also like to express their gratitude to several colleagues at Inco Limited (Sudbury), the Ontario Geological Survey and the University of Toronto for critical discussions that helped to improve the manuscript. P. Chevalier skillfully drafted most of the figures. Mrs. A. Branicky typed the manuscript.

References

Bayer. Geol. Landesamt (ed.), 1974: Die Forschungsbohrung Nördlingen 1973. Geologica Bavarica 72, München, Germany, 98 p.

Burrows, A.G. and H.C. Rickaby, 1929: Sudbury Basin Area. Ontario Dept. Mines, Ann. Rept. 38, pt. 3.

Card, K.D., 1978: Geology of the Sudbury-Manitoulin Area, Districts of Sudbury and Manitoulin; Ontario Geol. Surv., Report 166, 238 p. Accompanied by Map 2360, scale 1 inch to 2 miles or 1:126,720 and 4 charts.

Card, K.D. and R.W. Hutchinson, 1972: The Sudbury Structure: Its Regional Geological Setting. In: New Developments in Sudbury Geology (Guy-Bray, J.V., ed.). Geol. Assoc. Can., Spec. Paper 10, 67–78.

Cooke, H.C., 1946: Problems of Sudbury Geology. Geol. Surv. Can., Bull. 3, 77 p.

Daly, R.A., 1947: The Vredefort ring-structure of South Africa. J. Geol. 55, 125–145.

Dence, M.R., 1972: Meteorite Impact Craters and the Structure of the Sudbury Basin. In: New Developments in Sudbury Geology (Guy-Bray, J.V., ed.) Geol. Assoc. Can., Spec. Paper 10, 7–18.

Dietz, R.S., 1962: Sudbury Structure as an Astrobleme. Trans. Amer. Geophys. Union 43, 445–446.

Dietz, R.S., 1964: Sudbury as an Astrobleme, J. Geol., 72, 412–434.

Dietz, R.S., 1972: Sudbury Astrobleme, splash emplaced Sublayer and possible cosmogenic ores. In: New Developments in Sudbury Geology (Guy-Bray, J.V., ed.). Geol. Assoc. Can., Spec. Paper 10, 29–40.

Dressler, B.O., 1970: Die Beanspruchung der präkambrischen Gesteine in der Kryptoexplosionsstruktur von Manicouagan in der Provinz Quebec, Canada. Doctoral Thesis, University of Munich, Germany.

Dressler, B.O., 1984: The Effects of the Sudbury Event and the Intrusion of the Sudbury Igneous Complex on the Footwall Rocks of the Sudbury Structure. In: The Geology and Ore Deposits of the Sudbury Structure (Pye, E.G., A.J. Naldrett and P.E. Giblin, eds.). Ontario Geol. Surv., Spec. Vol. 1, 97–136.

Dressler, B., G. Graup und K. Matzke, 1969: Die Gesteine des kristallinen Grundgebirges im Nördlinger Ries. Geologica Bavarica 61, 201–228.

Dupuis, L., R.E.S. Whitehead and J.F. Davies, 1982: Evidence for a Genetic Link between Sudbury Breccias and Fenite Breccias, Can. J. Earth Sci. 19, 1174–1184.

Fairbairn, H.W., P.M Hurley, K.D. Card and C.J. Knight, 1969: Correlation of Radiometric Ages of Nipissing diabase and Huronian metasediments with Proterozoic Orogenic Events in Ontario. Can. J. Earth Sci. 6, 489–497.

Fairbairn, H.W. and G.M Robson, 1942: Breccia at Sudbury. Ontario. J. Geol. 50, 1–33.

Fairbairn, H.W. and G.M. Robson, 1943: Breccia at Sudbury. Ontario Dept. Mines, Annual Report for 1941, 50, Part 6, 18–33.

Frarey, M.J., W.D. Loverbridge and R.W. Sullivan, 1982: A U-Pb Zircon Age for the Creighton Granite, Ontario. In: Rb-Sr and U-Pb Isotopic Age Studies, Report 5; in Current Research, Part C, Geol. Surv. Can., Paper 82–16, 129–132.

French, B.M., 1968: Sudbury Structure, Ontario: some petrographic evidence for an origin by meteorite impact. In: Shock Metamorphism of Natural Materials (French, B.M. and N.M Short eds.) Mono Book, Baltimore, 383–412.

French, B.M., 1970: Possible Relations between Meteorite Impact and Igneous Petrogenesis, as indicated by the Sudbury Structure, Ontario, Canada. Bull. Volc. 34, 466–517.

French, B.M., 1972: Shock-Metamorphic Features in the Sudbury Structure, Ontario: A Review. In: New Developments in Sudbury Geology (Guy-Bray, J.V., ed.). Geol. Assoc. Can., Spec. Paper 10, 19–28.

Gasparrini, E. and A.J. Naldrett, 1972: Magnetite and Ilmenite in the Sudbury Nickel Irruptive. Econ. Geol. 67, 605–621.

Gibbins, W.A. and R.H. McNutt, 1975: The Age of the Sudbury Nickel Irruptive and the Murray Granite. Can. J. Earth Sci. 12, 1970–1989.

Gupta, V.K., F.S. Grant and K.D. Card, 1984: Gravity and Magnetic Characteristics of the Sudbury Structure. In: The Geology and Ore Deposits of the Sudbury Structure (Pye, E.G., A.J. Naldrett and P.E. Giblin, eds.). Ontario Geol. Surv., Spec. Vol. 1, 381–410.

Guy-Bray, J., and Geological Staff, 1966: Shatter Cones at Sudbury. J. Geol. 47, 243–245.

Hewins, R.H., 1971: The Petrology of some Marginal Rocks along the North Range of the Sudbury Irruptive. Ph.D. Thesis, University of Toronto.

Irvine, T.N., 1975: Crystallization Sequences of the Muskox Intrusion and other Layered Intrusions II. Origin of Chromitite Layers and similar Deposits of other Magmatic Ores. Geochim. Cosmochim. Acta 39, 991–1020.

James, C., D. Stöffler, A. Bischoff and W.U. Reimold, 1983: Reduction of SiO_2 to Si and Metallurgical Transformations in Al by Hypervelocity Impact of Al-Projectiles into Quartz Sand. Lunar Planet. Science XIV. Lunar and Planetary Institute, Houston, 347–348.

Krogh, T.E., D.W. Davis and F. Corfu, 1984: Pecise U-Pb Zircon and Baddeleyite Ages of Rocks from the Sudbury Area. In: Geology and Mineral Deposits of the Sudbury Structure (Pye, E.G. A.J. Naldrett and P.E. Giblin, eds.). Ontario Geol. Surv., Spec. Col. 1, 431–446.

Krogh, T.E., R.H. McNutt and G.L. Davis, 1982: Two high precision U-Pb zircon ages for the Sudbury Nickel Irruptive. Can. J. Earth Sci. 19, 723–748.

Kuo, H.Y. and J.H. Crocket, 1979: Rare Earth Elements in the Sudbury Nickel Irruptive: Comparison with Layered Gabbros and Implications for Nickel Irruptive Petrogenesis. Econ. Geol. 79, 590–605.

Lumbers, S.B., 1975: Geology of the Burwash Area, Districts of Nipissing, Parry Sound, and Sudbury. Ont. Div. Mines, GR. 116, 160 p. Accompanied by Map 2271, scale 1 inch to 2 miles.

Moore, E.S., 1930: Geological structure of the southwest portion of the Sudbury Basin. Trans. Can. Inst. Min. & Met. 33, 292–302.

Muir, T.L., 1982: Geology of the Morgan Lake – Nelson Lake area, District of Sudbury. Ontario Geol. Surv., Open File Report 5426, p. 1–203.

Muir, T.L., 1984: The Sudbury Structure: Considerations and Models for an endogenic origin. In: The Geology and Ore Deposits of the Sudbury Structure (Pye, E.G., A.J. Naldrett and P.E. Giblin, eds.). Ontario Geol. Surv., Spec. Vol. 1, 449–489.

Muir, T.L. and W.V. Peredery, 1984: The Onaping Formation. In: The Geology and Ore Deposits of the Sudbury Structure (Pye, E.G., A.J. Naldrett and P.E. Giblin, eds.). Ontario Geol. Surv., Spec. Vol. 1, 139–210.

Murtaugh, J.G., 1976: Manicouagan Impact Structure Area. Ministère des Richesses Naturelles, Quebec, DPV-432.

Offield, T.W. and H.A. Pohn, 1977: Deformation at the Decaturville Impact Structure, Missouri. In: Impact and Explosion Cratering (Roddy, D.J., R.O. Pepin and R.B. Merrill, eds.). Pergamon Press, New York, 321–341.

Naldrett, A.J., 1984: Mineralogy and Composition of the Sudbury Ores. In: Geology and Mineral Deposits of the Sudbury Structure (Pye, E.G., A.J. Naldrett and P.E. Giblin, eds.). Ontario Geol. Surv., Spec. Vol. 1, 309–325.

Naldrett, A.J., J.G. Bray, E.L. Gasparrini, T. Podolski and J.C. Rucklidge, 1970: Cryptic Variation and the Petrology of the Sudbury Nickel Irruptive. Econ. Geol. 65, 122–155.

Naldrett, A.J. and R.H. Hewins, 1984: The Main Mass of the Sudbury Igneous Complex. In: The Geology and Ore Deposit of the Sudbury Structure (Pye, E.G., A.J. Naldrett and P.E. Giblin, eds.)Ontario Geol. Surv., Spec. Vol. 1,235–251.

Naldrett, A.J., R.H. Hewins, B.O. Dressler and B.V. Rao, 1984: The Sublayer of the Sudbury Igneous Complex. In: The Geology and Ore Deposits of the Sudbury Structure (Pye, E.G., A.J. Naldrett and P.E. Giblin, eds.). Ontario Geol. Surv., Spec. Vol. 1, 254–274.

Naldrett, A.J., R.H. Hewins and L. Greenman, 1972: The Main Irruptive and the Sublayer at Sudbury, Ontario. Intern. Geol. Congress, 24th, Montreal, Proceedings Section 4, 206–214.

Naldrett, A.J. and G. Kullerud, 1967: A Study of the Strathcona Mine and its Bearing on the Origin of the Nickel-Copper Ores of the Sudbury District, Ontario. J. Petrol. 8, 453–531.

Naldrett, A.J. and A.J. MacDonald, 1980: Tectonic Settings of some Ni-Cu Sulfide Ores: Their Importance in Genesis and Exploration. Geol Assoc. Can., Spec. Paper 20, 633–657.

Pattison, E.F., 1979: The Sudbury Sublayer: Its Characteristics and Relationships with the Main Mass of the Sudbury Irruptive. Can. Mineralogist 17, 257–274.

Peredery, W.V., 1972a: Chemistry of fluidal glasses and melt bodies in the Onaping Formation. In: New developments in Sudbury geology (Guy-Bray, J.V. ed.). Geol. Assoc. Can., Spec. Paper 10, 49–59.

Peredery, W.V., 1972b: The origin of rocks at the base of the Onaping Formation, Sudbury, Ontario. Ph.D. Thesis, University of Toronto, 366 p.

Peredery, W.V. and G.G. Morrison, 1984: Discussion of the Origin of the Sudbury Structure. In: The Geology and Ore Deposits of the Sudbury Structure (Pye, E.G., A.J. Naldrett and P.E. Giblin eds.). Ontario Geol. Surv., Spec. Vol. 1, 491—511.

Peredery, W.V. and A.J. Naldrett, 1975: Petrology of the Upper Irruptive Rocks, Sudbury, Ontario. Econ. Geol. 70, 164—175.

Pye, E.G., A.J. Naldrett and P.E. Giblin (eds.), 1984: The Geology and Ore Deposits of the Sudbury Structure. Ontario Geol. Surv., Spec. Vol. 1, 603 p.

Rao, B.V., A.J. Naldrett, N.M. Evensen and B.O. Dressler, 1983: Contamination and genesis of the Sudbury Ores, Grant 146. In: Geoscience Research Grant Program, Summary of Research 1982—1983 (Pye, E.G., ed.). Ontario Geol. Surv., Miscellaneous Paper 113, 139—151.

Robertson, P.B. and R.A.F. Grieve, 1977: Shock Attenuation at Terrestrial Impact Structures. In: Impact and Explosion Cratering (Roddy, D.J., R.O. Pepin, and R.B. Merril, eds.). Pergamon Press, New York, 687—702.

Shoemaker, E.M., 1963: Impact mechanics at Meteor Crater, Arizona. In: The Solar System, Vol. 4 (Kuiper, G.P. ed.). University of Chicago Press, Chicago, 301—336.

Speers, E.C., 1957: The Age Relationship and Origin of Common Sudbury Breccia. J. Geol. 65, 497—514.

Stevenson, J.S., 1963: The upper contact phase of the Sudbury micropegmatite. Can. Mineralogist 7, 413—419.

Stevenson, J.S., 1972: The Onaping Ash-Flow Sheet, Sudbury, Ontario. In: New Developments in Sudbury Geology (Guy-Bray, J.V., ed.). Geol. Assoc. Can., Spec. Paper 10, 41—48.

Stevenson, J.S. and L.S. Stevenson, 1980: Sudbury, Ontario, and the meteorite theory. Geoscience Canada 7, 3, 103—108.

Stöffler, D., U. Ewald, R. Ostertag and W.H. Reimold, 1977: Research Drilling Nördlingen 1973 (Ries): Composition and Textures of Polymict Impact Breccias. Geologica Bavarica 75, 163—190.

Souch, B.E., T. Podolsky and Geological Staff, 1969: The Sulfide Ores of Sudbury: Their Particular Relation to a Distinctive Inclusion-Bearing Facies of the Nickel Irruptive. Econ. Geol., Monograph 4, 252—261.

Thomson, J.E., 1956: Geology of the Sudbury Basin. Ontario Dept. of Mines Annual Report 65, Pt. 3, 1—56.

Van Schmus, W.R., 1965: The Geochronology of the Blind River-Bruce Mines Area, Ontario, Canada. J. Geol. 73, 775—780.

Wilshire, H.G., 1971: Pseudotachylites from the Vredefort Ring, South Africa. J. Geol. 79, 195—206.

Wood, D.A., 1980: The application of a Th-Hf-Ta diagram to problems of tectonomagmatic classification and to establishing the nature of crustal contamination of basaltic lavas of the British Tertiary volcanic province. Earth Planet. Sci. Lett. 50, 11—30.

Yates, A.B., 1948: Properties of the International Nickel Company of Canada. In: Structural Geology of Canadian Ore Deposits. Canadian Inst. Mining and Metallurgy, 598—617.

Continuous Deposits of the Ries Crater, Germany: Sedimentological and Micropaleontological Investigations of NASA Drill Core D

Rolf Ostertag
Fraunhofer-Institut ISC, Neunerplatz 2, D-8700 Würzburg

Wolfgang Gasse
Geologisch-Paläontologisches Institut und Museum, Corrensstraße 24, D-4400 Münster

Dedicated to Prof. W. v. Engelhardt on the occasion of his 75th birthday

Key Words

Ries crater
Bunte Breccia
Continuous deposits
Impact crater
Impact ejecta
Ejecta emplacement
Grain size analyses
Clast size analyses
Clast provenance studies
Shock metamorphism
Microfossils

Abstract

NASA drill core D was taken about 25 km (2 crater radii) south of the center of the Ries crater in Southern Germany at $10°27'02''E/48°40'44''N$. It penetrated 52.3 m of the "Bunte Breccia" continuous deposits of the Ries crater, and 2.7 m of autochthonous Tertiary "Obere Süßwassermolasse" sediments, terminating in Tertiary "Obere Meeresmolasse" sediments at a depth of 62.7 m. The Bunte Breccia consists of lithic fragments derived from the crater cavity (crystalline basement, Triassic and Jurassic sediments) and from local material outside the actual crater rim (Tertiary freshwater and marine sediments). These clasts are embedded in four different types of a fine-grained groundmass made up of sands and clays. More than 4500 clasts > 2 mm in the Bunte Breccia were assigned to their stratigraphic provenance. Clasts 2–28 mm consist predominantly of crater derived material, while clasts 28–200 mm and > 200 mm are dominated by lithologies derived from outside the crater cavity (local material). The population of crater derived clasts of all size fractions is dominated by lithic clasts from the uppermost stratigraphic levels. A comparison of grain size, heavy mineral, carbonate content, and micropaleontological data of the groundmass with those of locally derived lithic clasts revealed that approximately 90 vol. % of the groundmass of the polymict breccia (grain size < 2 mm) consist of locally derived clastic sediments and only about 10 vol. % are crater derived. The breccia emplacement was a highly turbulent process which involved stripping of the local substrate as well as vertical and horizontal transport and intimate mixing of crater and locally derived materials. The high amount of locally derived material is predicted by the hypothesis of secondary cratering and associated mass wasting and is not compatible with a roll-and-glide mechanism of ejecta transport.

1. Introduction

Most of the samples from other planetary bodies such as lunar rocks and meteorites are impact breccias and their interpretation depends heavily on detailed knowledge of the source of individual components and their distribution in the impact deposits. By far the largest amount of impact-ejected material is located outside the actural crater rim, forming the continuous deposits of large impact craters.

Consequently, most of the extraterrestrial samples should be derived from outer impact formations and detailed investigation of terrestrial analogs of these breccias is crucial for any valid interpretation of meteorites and lunar rocks and their components. The Bunte Breccia of the Ries crater is the only example of a well-preserved continuous deposit of a large terrestrial impact crater. A very fortunate geologic and paleogeographic situation in the Ries area prior to the impact 15 million years ago allows to discriminate different breccia components and to assign these to their primary stratigraphic position. Moreover, it is even possible to distinguish between breccia components derived from the crater cavity and those which were incorporated into the deposits from the local substrate outside the actual crater area (Hüttner, 1961; Schneider, 1971). The amount of this locally derived material in the Bunte Breccia deposits is the key evidence to decide which transport mechanism leads to the formation of the continuous deposits of large impact craters, a nonballistic roll-and-glide mechanism (Chao, 1976) or a ballistic transport of ejecta with secondary cratering and associated mass wasting (Oberbeck, 1975; Morrison and Oberbeck, 1978). The model of ballistic transport predicts a large amount of locally derived material which even increases downrange, while nonballistically transported impact deposits would primarily consist of crater derived material with only a small basal mixing zone where components of the country rock might be incorporated.

This paper presents the results of our clast size, clast distribution, and clast provenance studies as well as the results of matrix studies based on grain size, carbonate content, and heavy minerals of a drill core taken in the Bunte Breccia continuous deposits of the Ries crater. In addition to the sedimentological investigations it appeared promising to include micropaleontological studies of clasts from the locally derived Tertiary material which are incorporated in the breccia deposits. The latter studies were expected to further characterize locally derived materials with respect to their stratigraphic position prior to the impact. Our major objectives were to determine the ratio of locally vs. crater derived material in the Bunte Breccia and to investigate the vertical distribution of crater and locally derived materials from different stratigraphic levels in these impact deposits. The drill core investigated in this study was part of an extensive drilling program performed by NASA in the Bunte Breccia SW of the Ries crater. Analyses of all nine drill cores and a comprehensive discussion of the results are given by Hörz et al., (1983). The present report emphasizes additional insight into the Bunte Breccia deposits via laboratory studies that are unique to this specific core.

2. Paleogeographic Situation Prior to the Impact

The Ries target consisted of a sequence of about 600 m of sedimentary rocks covering crystalline basement rocks of Variscan age (Schmidt-Kaler, 1969a; Köhler et al., 1981) The sedimentary strata capping the crystalline basement consist of Permian and Lower Triassic rocks of unknown thickness, and a well-defined sequence of Upper Triassic (Keuper), and Jurassic (Lias, Dogger, Malm) sediments. The Upper Jurassic limestones

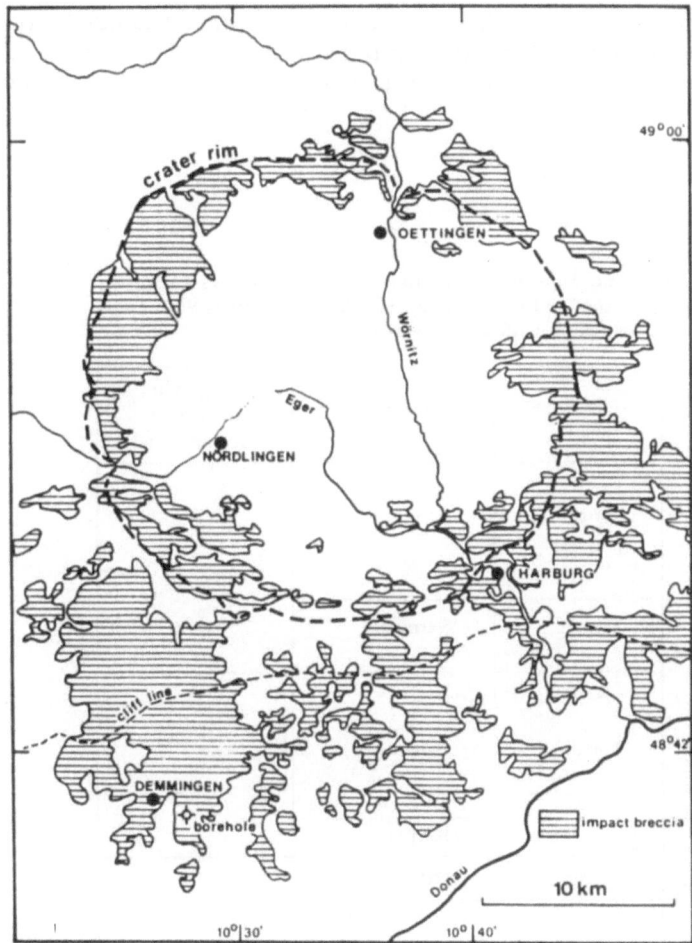

Fig. 1
Distribution of the Bunte Breccia continuous deposits of the Ries crater.
Drill location D is shown in the SW about 6 km south of the Miocene cliff
line (map modified after Chao et al., 1978).

were covered by erosional remnants of Tertiary freshwater sediments. The target rocks gently dip toward SE. At a distance of 6 km south of today's crater rim, a Miocene cliff line running E-W (Gall, 1974) marked the Northern shore of the South German Molasse Sea (Fig. 1). The marine sediments of the "Obere Meeresmolasse, OMM" were confined to the south of the cliff line where they were covered by freshwater sands and clays of the "Obere Süßwassermolasse, OSM". It is generally agreed that the OSM sediments capping the Upper Jurassic in the crater area north of the cliff line were eroded to a large extent prior to the impact 15 million years ago (Hüttner, 1961; Birzer, 1969; Bolten and Müller, 1969). It is therefore justified to presume that only rocks of Mesozoic and older age in the Bunte Breccia at drill site D are crater derived.

Consequently, all Tertiary materials in the Bunte Breccia deposits are considered "locally derived" from outside the crater cavity. Table I gives a time scale and summarizes the stratigraphic sequence of Tertiary sediments south of the cliff line which are incorporated in the Bunte Breccia deposits.

Fortunately, the target rocks and the local materials are very different in their lithologies, color, grain size, and mineral content so that even fairly small clasts may be attributed to their source without major uncertainties. These fortunate circumstances are the basis for our investigations.

Table 1: Time scale and stratisgraphic sequence of Tertiary in the Paratethys (modified after Hagn 1981), revised stages according to Steiniger et al., (1976); OSM: Obere Süßwassermolasse, OMM: Qbere Meeresmolasse, USM: Untere Süßwassermolasse.

mio. y.	series	stages		stratigraphic sequence of local material
		obsolete	revised	
10	MIOCENE — UPPER	Pont	Pannon	
		Sarmat	Sarmat	
				– – – – ? – – – –
15	MIOCENE — MIDDLE	Torton	Baden	◄— Ries event —►
				OSM
			Karpat	
		Helvet	Ottnang	
20	MIOCENE — LOWER	Burdigal	Eggenburg	OMM
		Aquitan	Eger	USM

3. Experimental

Drill core D (formerly called drill core # 2; Hörz et al., 1977), was taken on a hill near the village of Demmingen about two crater radii (25 km) south of the crater center and 6.5 km south of the cliff line. The drilling location ($10°27'02''$E/$48°40'44''$N, 534 m above sea level) was placed in a huge clast of OMM sand (Fig. 2). Standard techniques were employed for sample preparation. 43 specimens of clasts > 200 mm in size and 16 specimens of the polymict breccia (< 200 mm) (dry weight 100 g each) were disaggregated in H_2O_2 and nine grain size fractions were obtained by wet sieving (1, 0.5–1, 0.25–0.5, 0.125–0.25, and 0.063–0.125 mm) and sedimentation techniques (Atterberg cylinder) (20–63, 6.3–20, 2–6.3, and $< 2\,\mu$m). Additional grain size analyses with a semiautomatic Zeiss linear analyser were made in thin section and the clast size distribution was studied on 1:1 photographs using a Zeiss TGZ 3 grain size analyser. Grain size parameters were calculated according to Folk and Ward (1957). Clasts were divided into three size classes (2–28 mm, 28–200 mm, and > 200 mm) for technical reasons (limited range of the TGZ 3 grain size analyser). This division also gave valuable insight into the relationship between clast size and source lithology. Clast lithologies were determined with the naked eye aided by a hand lens, supported by grain size and heavy mineral data and carbonate content, and modal and textural analyses of selected clasts in thin section. Heavy minerals were separated from the 0.063–0.125 mm grain size fractions with bromoform (d = 2.86 g/cm^3). The nature of the different heavy minerals was determined by optical methods and by single crystal Xray techniques using a Gandolfi camera. The carbonate content of powdered samples was determined gasometrically and was calculated as $CaCO_3$. The determination of the microfossil content was carried out under the microscope and the SEM basically considering the grain size fraction > 0.2 mm.

4. Results

Macroscopic Core Description

The drill core was 10 cm in diameter and 62.7 m long. It penetrated autochthonous Tertiary freshwater limestone ("Lepolithkalk") at 52.3 m and terminated in a light green OMM silt containing abundant calcareous nodules (Fig. 2). The autochthonous rock strata are in an undisturbed position as indicated by a thin horizontal bed of mollusc shell detritus in the silt, and they represent the normal stratigraphic sequence of Molasse sediments in this area (Kiderlen, 1931; Hüttner, 1961; Bolten and Müller, 1969). The contact between the Bunte Breccia and the country rock at 52.3 m is extremely sharp and well-defined (Fig. 3). A basal mixing zone (Chao, 1977) is not present. No relics of soil or weathering products are observed at the contact which indicates considerable erosion of the pre-Ries land surface during deposition of the Bunte Breccia. The mild and shallow brecciation of the friable autochthonous freshwater limestone at the contact reveals at best a minor vertical stress component during ejecta emplacement.
The Bunte Breccia section of the drill core is a highly chaotic arrangement of randomly distributed lithic clasts of different size, shape, texture, and stratigraphic provenance embedded in or intercalated with different types of a sand- and clay rich groundmass. Lithic clasts larger than 200 mm each account for 63 % or roughly two thirds of the Bunte Breccia section, 37 % consist of a polymict breccia with a grain size of individual clasts of < 200 mm. The thickness of individual polymict breccia sections ranges from a few cm to several m (Fig. 2).

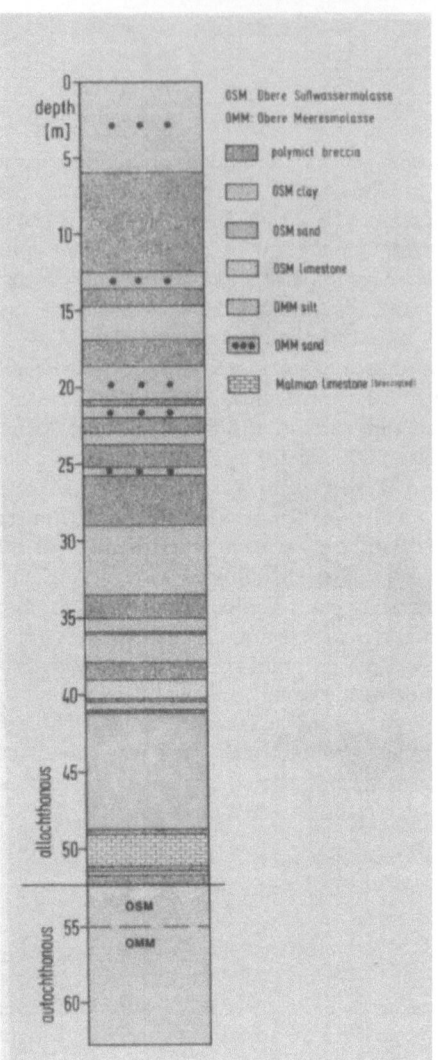

Fig. 2
Profile of core D; lithic fragments larger than 200 mm, and polymict breccia (grain size < 200 mm) are shown separately.

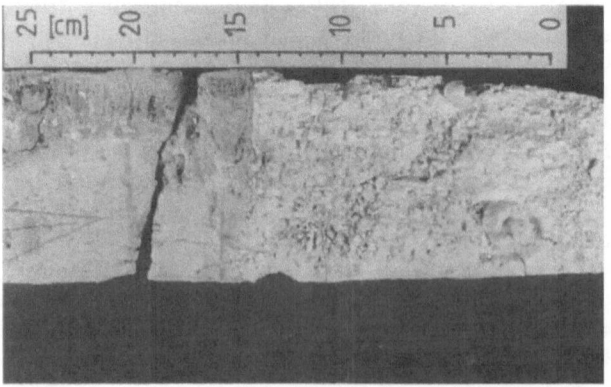

Fig. 3
Photograph showing the sharp contact between the Bunte Breccia (left) and the autochthonous OSM limestone (right). The limestone is only mildly brecciated, no basal mixing zone is present; weathering products on top of the OSM limestone have been eroded during breccia deposition.

The clast sizes in the Bunte Breccia of drill core D range from micron-sized mineral clasts to lithic clasts several m in dimension and the breccia components are distributed randomly. The clast orientation is random as well, except for one cm-sized core section where elongated clay clasts are oriented horizontally. In contrast, microscopic observations in thin sections show a preferred orientation of mica and other sheet-like, tabular or prismatic minerals. This preferred orientation, however, is heavily disturbed in the vicinity of lithic fragments. The overall random orientation of all lithic clasts is a general feature in the Ries continuous deposits and it was observed in all drill cores of the entire drilling program (Hörz et al., 1977, 1983).

No primary stratigraphic relations have been preserved among crater derived components which indicates a highly turbulent breccia emplacement mechanism. Locally derived Tertiary clasts, however, reflect an inversion of their normal stratigraphic position: The older marine sand clasts lie topographically higher than the younger marine silts which in turn rest on top of freshwater sand and limestone clasts (Fig. 2). The vertical uplift of the OMM sand in the first few core meters is at least 44 m as compared to autochthonous OMM sand outcropping in close vicinity of the drill location. OMM silts were uplifted by more than 17 m with respect to the autochthonous silt in the drill core, and the OSM limestone clast is located 12 m above its autochthonous counterpart. This inverted stratigraphic position of locally derived material is unique to drill core D.

About 90 vol. % of the polymict breccia (grain size < 200 mm) consist of a groundmass with a grain size of individual components of less than 2 mm. Clasts (> 2 mm by definition) floating in this groundmass are fragments of crystalline basement rocks (very rare), Triassic, Lower Jurassic (Lias and Dogger combined), and Upper Jurassic as well as Tertiary sand, sandstone, marl, limestone, silt, shale, and clay. Minor but genetically important components of this breccia are Upper Jurassic limestone pebbles with pholade bore holes which indicate derivation from the Miocene cliff line, detritus of oyster shells (Tertiary, OMM), and lignite pieces (Tertiary, OSM). Reddish or brown lateritic clays with abundant limonite are remnants of weathering products from the pre-impact land surface.

On the basis of color and the clay/sand ratios four clearly discernible types of groundmass (grain size < 2 mm) can be distinguished, each appearing rather homogeneous: a light brown and a dark brown type, both with a clay/sand ratio of > 1, and a tan as well as a grey type which display clay/sand ratios of < 1. A close relationship to the tan OMM sand and the grey OSM sand is indicated for the sand-rich types. The tan sand-rich and the light brown clay-rich types are dominant in the upper 22 m of the drill core where also the tan sand clasts prevail, while the grey sand-rich and the dark brown clay-rich types are more abundant in the lower 30 m of the Bunte Breccia section where the grey OSM sands occur. The different groundmass types are clearly separated from each other by sharp contacts. Even inclusions of one breccia type within another can be observed. Monomict brecciated limestone clasts are included in the polymict breccia sections as well. Thus a highly complex texture of breccias within breccias already results from one single impact as noted before by Hörz et al., (1977, 1983).

Clast Studies

Crystalline basement rocks and shock effects: Lithic fragments derived from the crystalline basement of the Ries crater (Graup, 1978) are rare among the Bunte Breccia clasts, and are generally small (< 5 mm) and well-rounded. Seventeen clasts were studied in thin section, 14 of which were biotite-plagioclase gneisses, one quartz-bearing amphibolite, one amphibole-bearing biotite-plagioclase gneiss, and one quartz-feldspar intergrowth.

Table II: Shock deformation of minerals in crystalline basement rocks in the Bunte Breccia deposits of drill core D, shock classification according to Stöffler (1971).

mineral	shock effect	shock stage
quartz	planar fractures, several systems of non-decorated planar elements	I
feldspars	planar elements isotropic lamellae	I
biotite	kink bands partially opaque	I (-II)
hornblende	heavily fractured	I
garnet	fractured	I
apatite	fractured	I
zircon	fractured	I

Two of the samples are unshocked, the others display shock features indicative of shock stages I and II (Stöffler, 1971), i. e. the maximum shock pressure did not exceed 40 GPa (Table II). It has to be noted that only rock fragments from the crystalline basement do show shock effects at all, all other lithic and mineral clasts in the Bunte Breccia are unshocked. The results obtained here are perfectly consistent with those obtained by Hörz and Banholzer (1980) on the crystalline basement rocks in the other drill cores and those of Graup (1978) for many additional Bunte Breccia locations.

Triassic and Jurassic clasts: Triassic (Keuper) rock clasts are predominantly red or purple clays and shales. Most of them are elongated and twisted, but some appear to be virtually unaffected by the transport. Small and well-rounded sandstone clasts which might be Triassic were found in thin sections. Grey or black shales and clays are attributed to the Lower Jurassic (Lias and Dogger), while Upper Jurassic clasts consist of light grey marly limestone and white limestone. All of these limestone clasts are angular, some are brecciated. One large Upper Jurassic limestone clast (2.1 m core length) is located at 49—51 m (Fig. 2).

Tertiary (OSM and OMM): The Tertiary freshwater sediments (OSM) in the Bunte Breccia of drill core D consist of rare light red or dark green clays, and more frequently of plastically deformed green and brown banded clays (3.3 m total core length), and a marly limestone clast ("Lepolithkalk") (1.2 m). The limestone is identical to the autochthonous limestone encountered at 53.2 m. OSM sands are medium-grained, micaceous, and show inclusions of marly nodules and lignite. Depending on pre-Ries weathering they are either yellow (0.8 m), containing limonite or grey, (6.7 m) containing pyrite. In the grey sands gastropode shell detritus is frequent. The large OSM limestone and sand clasts are concentrated in the lower half of the Bunte Breccia section of the drill core (Fig. 2). The OMM clasts consist of tan, fine-grained glauconitic sands with whitish, soft carbonate nodules (10.7 m), and light green silts sometimes with light brown intercalations (7.9 m). These sediments contain foraminifera and oyster shell detritus. OMM sands and silts are the dominant large clasts within the upper 40 m of drill core D. The green silts are texturally similar but not identical to the silts in the autochthonous part of the drill core.

Clast Size Analyses

Exact clast size analyses were performed on photographs of 12 breccia sections represent-
ing a total area of 3420 cm^2 for the clast size fraction 2-28 mm. The size of 1169 clasts
was measured and the statistical parameters were calculated according to Folk and Ward
(1957) (Fig. 4). The graphic mean (GM) is consistently of the order of −4 Phi (= 16 mm),
and sorting (IGSD) is very poor. The skewness (IGS) of the grain size distribution shows
excess of fine-grained material, and the kurtosis (K) varies between leptokurtic and
platycurtic. The statistical parameters of the clast size distribution in the 2-28 mm clast
size fraction are remarkably constant throughout different polymict breccia sections of
the drill core.

The mean grain size of the 28-200 mm clasts is of the order of 60 mm. The other grain
size parameters of the clasts size fractions of 28—200 mm and of clasts 200 mm have not
been calculated, because the clast sizes vary extremely (e. g., between 200 and 7500 mm)
and the number of clasts was too small to yield any statistically significant results.

Clast Provenance and Frequency Statistics

More than 4500 clasts > 2 mm were assigned to their primary stratigraphic position inside
or outside the crater area. Supporting evidence for their stratigraphic assignment came
from grain size data, heavy mineral analyses, carbonate content, macrofossil and the
microfossil populations, and microscopic investigations in thin section.

The 2-28 mm clast size fraction (4403 clasts counted) is dominated by crater derived
materials (Table III). Clasts from the uppermost rock strata of the target (Upper Jurassic)
dominate the distribution, followed by clasts from the Lower Jurassic (Lias and Dogger).
Triassic and crystalline basement rocks combined contribute only about 10 % to this
clast population. Locally derived clasts are less frequent (16 %) than crater derived materials. Fig. 5 displays the number frequencies of 2-28 mm clasts in the polymict breccia
sections of drill core D. The clast population in the 28-200 mm fraction (90 clasts) is
already dominated by locally derived components. Among the crater derived material,

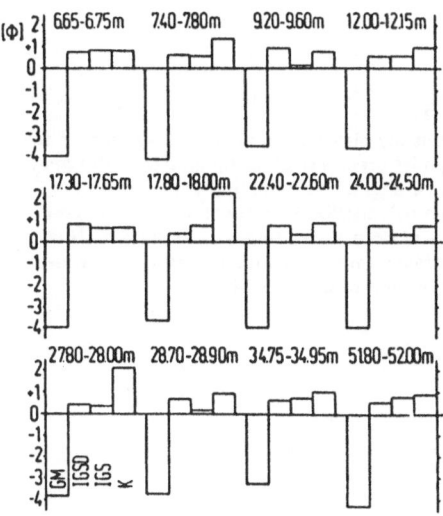

Fig. 4

Statistical parameters of clast size distribution
(2—28 mm). Note the similarities of the graphic
mean values of clasts from different depths, no
sorting of clasts is apparent in this vertical
section through the Bunte Breccia. GM: Graphic
Mean, IGS: Inclusive Graphic Skewness, IGSD:
Inclusive Graphic Standard Deviation, K: Kurtosis,
(Folk and Ward, 1957).

Table III: Clast population in drill core D (number frequencies).

clast size: source:	2—28 mm		28—200 mm		> 200 mm	
	No.	(%)	No.	(%)	No.	(%)
Tertiary	704	16.0	64	70.3	14	93.3
Upper Jurassic	1854	42.1	19	22.0	1	6.7
Lower Jurassic	1360	30.9	5	5.5		
Triassic	454	10.3	1	1.1		
Crystalline basement rocks	31	0.7	1	1.1		
Total	4403	100.0	90	100.0	15	100.0

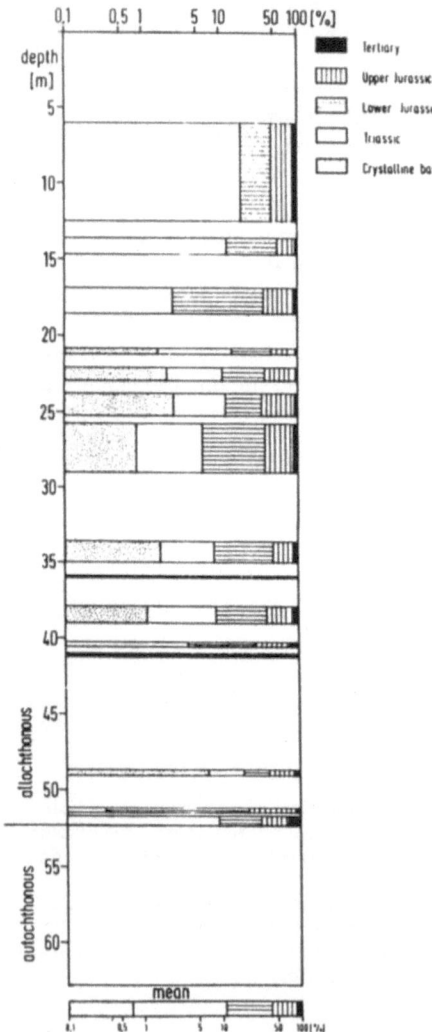

Fig. 5

Frequency distribution of clasts 2—28 mm in the polymict breccia sections (grain size < 200 mm) of drill core D. Clasts of all lithologies are homogeneously distributed throughout the drill core, clasts of deep seated target rocks are rare. Note the log scale on the horizontal bar. Bottom: average distribution of clasts 2—28 mm.

Upper Jurassic clasts again are much more frequent than fragments of Lower Jurassic rock strata. Triassic rocks and clasts from the crystalline basement are exceedingly rare in the 28-200 mm size fraction (Table III). In the clast size fraction > 200 mm only one 2.1 m Upper Jurassic limestone clast presumably originated from the crater area while all other lithic fragments (OMM sands and silts, OSM limestone, sand, and clay) are considered locally derived. The clast size and provenance studies reveal that already at a distance of 2 crater radii from the crater center only the smallest clasts are dominated by crater derived material, whereas large clasts are more likely to be locally derived. Furthermore, crater derived clasts of all sizes are dominated by the uppermost stratigraphic horizons of the target.

Sedimentological Investigations of Clasts and Breccia Groundmass (Grain Size < 2 mm) and Autochthonous Tertiary

The groundmass (< 2 mm grain size by definition) was analyzed in terms of grain size, $CaCO_3$ content, heavy mineral content, and fossil content. For comparison, the same properties were analyzed in clasts > 200 mm and in samples from the authochthonous Tertiary. The goal of this comparison was to determine the contribution of the locally derived Tertiary material in the grain size fraction < 2 mm of the polymict breccia.

Grain Size Analyses

The grain size distributions of representative samples of OSM and OMM sands (Fig. 6) show a mode between 2 and 3 Phi (0.125—0.25 mm) which is more prominent in the OSM sands than in the OMM sands. Two small layers of fine-grained marine sands, intermediate in grain size between OMM sand and silts, display a bimodal grain size distribution with one maximum at 2-3 Phi and one at > 4 Phi. The OMM silts have their mode at > 4 Phi (< 63 microns). The OSM clays are very fine grained with 50 % of their grains smaller than 2 μm.
The grain size distribution of the polymict breccia groundmass is bimodal displaying a first maximum in the size fraction 2-3 Phi, and a second maximum in the size fraction > 4 Phi (Fig. 6). Only a few exceptions to this rule exist. In some of the breccia sections a noticeable contribution of particles 1 mm is observed. These particles almost exlusively consist of Upper Jurassic limestone. The grain size data in the grain size fraction > 4 Phi are somewhat biased by desintegrated clay clasts but this contribution to the grain size distribution is only minor and does not account for the high amount of material in this size fraction.
A comparison of the grain size data of the Tertiary clastic sediment clasts and the polymict breccia groundmass reveals that the grain size pattern of the breccia groundmass (< 2 mm) can be derived by mixing locally derived sands and silts. The pronounced grain size maximum at 2-3 Phi in the breccia results from the contribution of Tertiary sand, while the second maximum at > 4 Phi is derived from OMM silts and probably OSM clays. In addition to the sieve analyses, grain size data were obtained by measuring thin sections of the polymict breccia and three representative samples of Tertiary sediments with a Zeiss linear analyser. The data obtained by this method included volume fractions and statistical parameters of grains 4 μm to 2 mm in diameter calculated according to Folk and Ward (1957), and the amount of "matrix" consisting of components < 4 μm in size. In addition, the frequency of contacts between discrete components > 4 μm and the matrix (< 4 μm) was determined (Table IV). The particles > 4 μm consist of mineral grains, limestone particles, crystalline rock clasts, clays, and shales. In contrast, particles

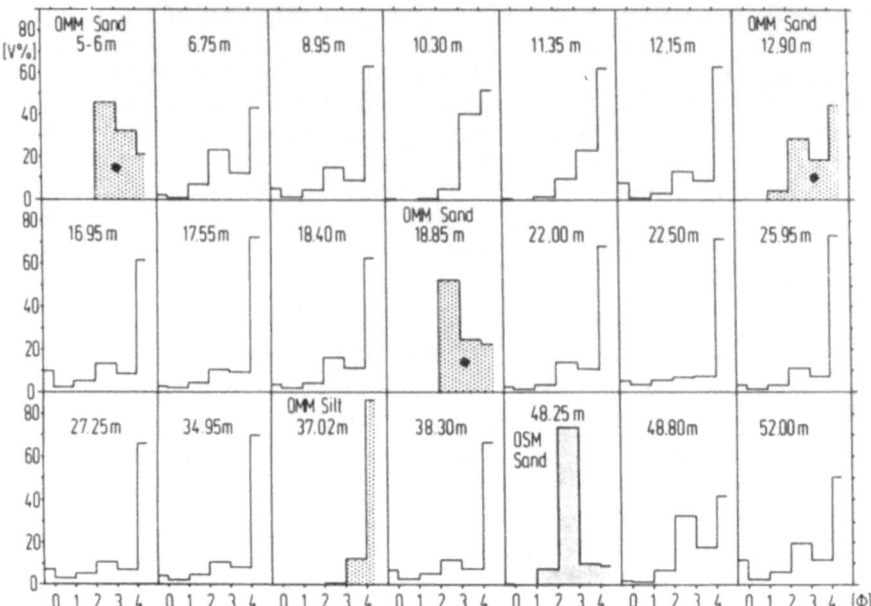

Fig. 6

Histograms of the grain size distribution of pure Tertiary clasts larger than 200 mm (stippled) and of the polymict breccia groundmass (grain size < 2 mm) (open).

Table IV: Linear analyses of Bunte Breccia and pure Tertiary in thin section; GM: Graphic Mean IGS: Inclusive Graphic Skewness, IGSD: Inclusive Graphic Standard Deviation, K: Kurtosis (Folk and Ward, 1957); V: volume percentage of components $< 4\,\mu m$; C: frequency of contacts between components $> 4\,\mu m$ components $< 4\,\mu m$.

sample (depth, m)	GM (Phi)	IGS (Phi)	IGSD (Phi)	K (Phi)	V (%)	C (%)
6.75	4.386	1.359	0.157	0.865	38.1	94.1
8.95	4.761	1.230	0.002	0.880	50.5	97.8
12.15	4.493	1.193	− 0.107	0.940	41.6	97.8
16.95	4.647	1.192	0.004	0.958	53.6	98.7
17.55	4.812	1.325	− 0.112	1.040	57.0	97.6
22.50	5.125	1.290	− 0.224	0.981	54.6	99.6
29.95	4.905	1.332	− 0.116	0.838	54.9	98.5
27 25	4.688	1.295	− 0.116	0.828	56.9	98.1
34.95	4.847	1.252	− 0.172	0.981	52.9	98.0
38.30	4.890	1.262	− 0.275	1.064	59.9	97.4
48.70	4.092	1.079	0.117	0.882	38.5	95.6
OMM sand						
5−6	3.883	0.865	0.228	1.037	35.9	83.6
OMM silt						
37.20	5.383	0.955	− 0.062	0.891	65.3	90.4
OSM sand						
45.30	4.319	1.202	− 0.124	0.860	41.8	87.4

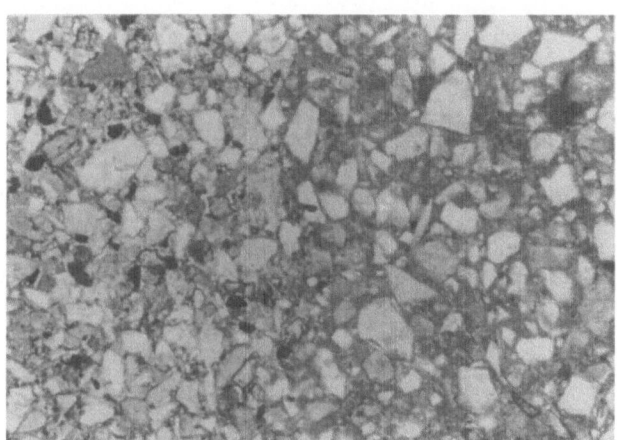

Fig. 7

Photomicrograph of a Bunte Breccia thin section (48.72 m); left: medium-grained micaceous OSM sand, right: polymict breccia groundmass. Note the differences in the amount of clay minerals. Transmitted light, one polar, fied width 3 mm.

$< 4 \mu m$ in general are kaolinite, montmorillonite, and mixed layers, as determined by Xray methods. The Graphic Means (GM) of the breccia sections are intermediate between the lower and upper values of the Tertiary sediments. The same holds for the percentage of material $< 4 \mu m$. These data are consistent with a mixture of locally derived Tertiary material for the groundmass (< 2 mm) of the polymict breccia. An admixture of OSM clay does not appear to be necessary and can even be excluded because the volume percentage of material $< 4 \mu m$ in the breccia is constantly below the respective value of the OMM silt. Most interestingly the components $> 4 \mu m$ are almost completely surrounded by clay minerals in the breccia which is significantly different from pure Tertiary clastic sediments (Table IV). This distinct difference in texture between an OSM sand clast and the breccia groundmass is shown in Fig. 7. An admixture of clay is unlikely to produce this texture as this would contradict the data listed in the preceeding column of Table IV. This particular texture of the Bunte Breccia section, i. e. nearly all minerals and rock clasts are floating in a clayey groundmass, seemingly requires a highly turbulent breccia emplacement mechanism.

Carbonate Content

Fig. 8 displays the carbonate content calculated as $CaCO_3$ of all clasts > 200 mm and that of autochthonous country rock and 21 polymict breccia sections which were free of clasts. The OMM sands at the top of the drill core are depleted in carbonate content to a depth of 4 m due to weathering. Below this depth the sands have a $CaCO_3$ content of 20—30 wt.%. The OMM silts display a more variable carbonate content ranging from 9 to 34 %. The yellow OSM sand is depleted in $CaCO_3$ relative to its grey counterpart. The grey sand shows a variable carbonate content (8—29 % $CaCO_3$) because of irregularily distributed marly nodules and gastropode shell detritus. The autochthonous bedrocks show a successive decrease in $CaCO_3$ with increasing depth except for a thin layer enriched in mollusc shell detritus at 59 m. In contrast to the variable $CaCO_3$ content of the above pure Tertiary lithologies, the carbonate content of the polymict breccia groundmass is remarkably constant. The data scatter between 13 and 20 % $CaCO_3$ only except for three samples in the 20-30 % range (Fig. 8). Evidently no carbonate transport was effective after deposition of the breccia. The carbonate contents of the Upper Jurassic and OSM

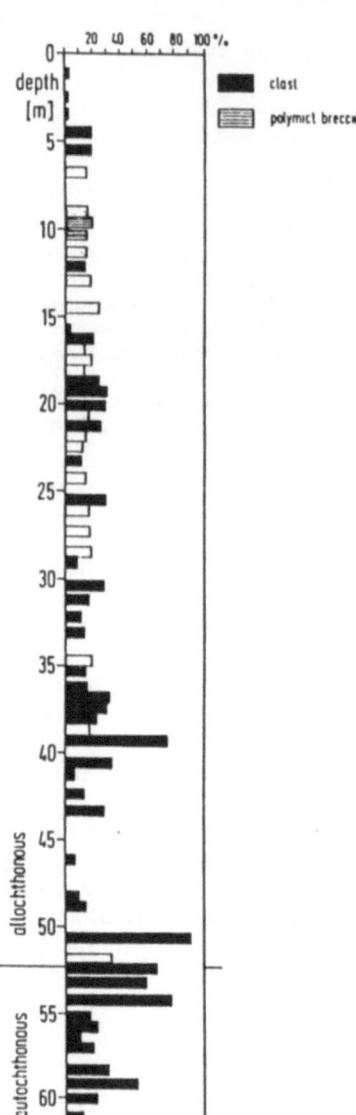

Fig. 8

CaCO$_3$-content of clasts > 200 mm
(black) and breccia groundmass (striped)
in drill core D. The carbonate content of
the breccia groundmass is remarkably
constant.

limestone clasts are not changed by secondary alteration and the polymict breccia sections adjacent to these clasts are not enriched in CaCO$_3$ as compared to the mean value of the polymict breccia groundmass. No basal mixing zone can be detected in the lowermost core section which directly overlays OSM limestone. In accordance with the grain size data, mixing of locally derived Tertiary sediments could produce the carbonate content of the polymict breccia groundmass.

Heavy Minerals

The most promising way to discriminate locally and crater derived stratigraphic units and to deduce their relative contributions to the polymict breccia groundmass (< 2 mm) is the absolute heavy mineral content and the heavy mineral population in comparison to potential source lithologies. Except for hornblende in amphibolites, heavy minerals in crystalline basement rocks are accessory phases only. Graup (1978) lists apatite, augite, sphene, zircon, rutile, epidote, orthite, clinozoisite, monazite, cordierite, and garnet to be the most prominent heavy minerals in the crystalline basement rocks in addition to hornblende. The total concentration of heavy minerals in the Bunte Breccia derived from the crystalline basement rocks appears neglegible because of the scarcity of these rocks in the continuous deposits outside today's crater rim. Major contributions of heavy minerals from finely comminuted crystalline material which may be there in disproportionate amounts relative to clasts are also unlikely to exist. Among the more than 17 000 mineral grains checked individually during the linear analyses of the polymict breccia and additional thin section investigations only one shocked quartz grain and one biotite with kink bands were found. This number should be considerably increased if large amounts of finely comminuted crystalline basement rocks were part of the polymict breccia groundmass.

The heavy mineral spectrum of Triassic and Jurassic sediments in the Ries area displays zircon along with tourmaline and minor contributions of garnet as its main constituents (Schnitzer, 1953; Schröder, 1962; Schmidt-Kaler, 1969b; Salger and Schmidt-Kaler, 1973, 1974). Mesozoic sandstones which could contain appreciable amounts of heavy minerals are rare among the clasts in the Bunte Breccia at drill site D. Therefore any major contribution of mesozoic sediments to the heavy mineral spectrum of the polymict breccia groundmass is not expected.

The heavy mineral separates of the Tertiary sediments are vastly dominated by garnet, epidote, and opaques such as pyrite and limonite. Garnet and epidote comprise more than 90 % of all transparent heavy minerals. Among the remaining < 10 % pumpellyite, rutile, staurolite, tourmaline, cyanite, sphene, apatite, hornblende, zircon, and clinozoisite were observed.

Fortunately, and despite possible secondary alterations of the heavy mineral spectra of OMM and OSM sediments, the garnet/epidote ratio may be used to separate these stratigraphic units. Limitations come from the considerable variability in the garnet/epidote ratio of OSM sediments where individual ratios may range from 2 to about 60 (Lemcke et al., 1953). In addition, selective weathering of garnet may change the garnet/epidote ratio towards lower values. Selective weathering of garnet is effective in samples depleted in carbonate content because garnet is more readily dissolved by acidic solutions than epidote (Schmeer, 1955). Nevertheless, a garnet/epidote ratio of > 1 in OSM sediments and a ratio < 1 in OMM sediments is generally maintained and thus allows the distinction between these stratigraphic units even as clasts in the Bunte Breccia. The absolute heavy mineral content in the grain size fraction 0.063—0.125 mm totals 3 wt.% in the OMM clasts, 13 wt.% in the OSM sand clasts, and 1 wt.% in the autochthonous OMM silt. Pyrite and limonite are included and account for the high percentages of heavy minerals in the OSM sands.

Fig. 9 summarizes the heavy mineral content of 12 core sections. The dominance of epidote in the OMM sediments and the dominance of garnet in the heavy mineral spectra of OSM sand is evident. The heavy mineral content of the OSM clays was too low to yield any statistically significant results. The average garnet/epidote ratio of OMM clasts in drill core D is 0.35. The yellow OSM sand which probably had been subjected to

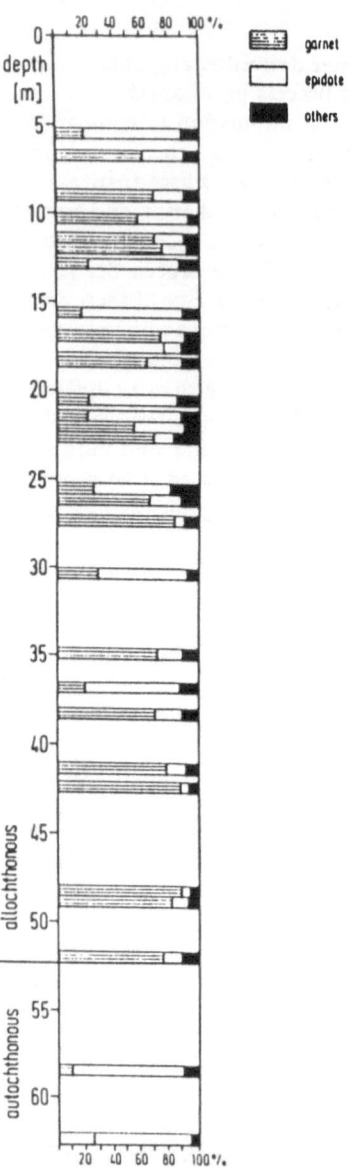

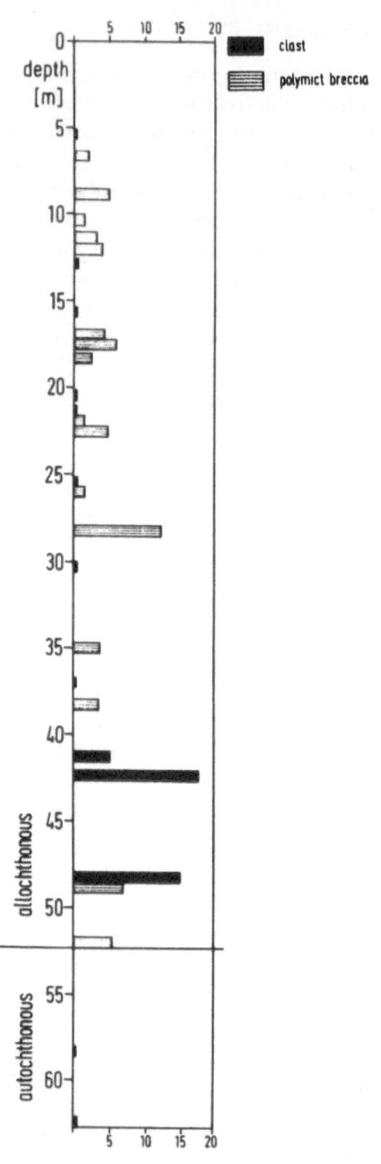

Fig. 9

Heavy mineral population in pure Tertiary clasts and polymict breccia groundmass in the grain size fraction 0.063–0.125 mm; only transparent heavy minerals were considered. Striped: garnet, open: epidote, black: others, see text.

Fig. 10

Garnet/epidote ratio in the grain size fraction 0.063–0.125 mm of pure Tertiary clasts and the breccia groundmass. The g/e ratio of the breccia groundmass (striped) is intermediate between those of pure OMM and OSM sediments (black).

weathering prior to the Ries impact yields a ratio of 5, while the grey OSM sand averages 16 (Fig. 10) The high absolute heavy mineral content and the high number of Tertiary clasts in the Bunte Breccia at drill site D finds its expression in the heavy mineral spectrum of the Bunte Breccia groundmass. The heavy mineral content in the grain size fraction 0.063–0.125 mm of the polymict breccia amounts to 3 wt.% including pyrite and limonite. The average transparent heavy mineral population in the Bunte Breccia displays 70 % garnet, 20 % epidote, 3 % tourmaline, 3 % staurolite, 2 % pumpellyite, and 2 % others such as zircon, rutile, and apatite (Fig. 9). The garnet/epidote ratios in the polymict breccia groundmass fall within the range given by the Tertiary clasts, they vary between 1.6 and 12 (Fig. 10).

The overall content of heavy minerals in the polymict breccia groundmass which might probably be derived from Mesozoic sediments and crystalline basement rocks (zircon, tourmaline, hornblende) does not exceed the relative frequency of these minerals in the Tertiary sediments. Therefore any major contribution of Mesozoic material to the groundmass of the polymict breccia can again be precluded. The heavy mineral content, the grain size data, and the carbonate content consistently and independently favor a derivation of the polymict breccia groundmass (grain size < 2 mm) from locally derived Tertiary clastic sediments.

Micropaleontological Investigations

Analyses of the microfossil content (benthonic foraminifers, Table V, and ostracodes) of the autochthonous Tertiary sediments and of the Tertiary clasts were performed to further characterize these sediments and their environmental conditions during formation. These data should provide additional insight in the nature of the source area(s) of the Tertiary clasts now residing in the Bunte Breccia.

The facies of the OMM sediments South of the cliff line range from marine in the fine-grained sands at the bottom to pliohaline in the silts at the top of the stratigraphic sequence (Schetelig, 1962). Only benthonic forams indicative of shallow water occur in the Tertiary marine sediments of the drill core. The depth of the sea in the area where the OMM sediments formed was less than 200 m as determined from the foram population (Gasse, 1981). This corresponds closely with the proximity of the Miocene cliff line (Fig. 1). A comparison with benthonic foram assemblages reported from the Miocene Molasse sea and the Parathetys (Brestenska, 1974; Cicha et al., 1971, 1973; Hagn, 1981) yielded a lower Miocene age (middle to upper Ottnang; Table I) for all these sediments. This age is consistent with the biostratigraphic assignment of OMM sediments to the lower Miocene by Schetelig (1962) and Gall (1971) in areas close to drill site D. Facies-independent index fossils are missing, however, and an exact assignment of the OMM clasts to stratigraphic subunits proved to be difficult.

As a consequence of the variability of environments during sedimentation of the OMM, differences exist between microfossil populations of individual sand and silt clasts in the Bunte Breccia (Table V). Interestingly, the microfossil contents of the OMM silt clasts and their autochthonous counterparts at site D differ as well. The autochthonous OMM and the OMM clasts differ in their ostracode content and the number of foram genera and species. Ostracodes were only observed in the bed rock and none were found in the displaced OMM silt clasts. It is unlikely, however, that the friable ostracode shells were depleted quantitatively by comminution during ejecta transport, because many primary sedimentary textures have been preserved in these large unconsolidated clasts. Therefore the differences in the ostracode contents of these two OMM silts must be a primary feature, indicating differences in the exact source area(s) of these sediments.

Table V: Fossil content of pure Tertiary clasts and autochthonous bedrock of drill core D, and the interpretation of environmental conditions during deposition of OMM sediments based on the occurence and diversity of foram genera and species.

		Species	Systematic description by
Character of wall	agglutinated	*Thurammina papillata* Brady, 1879	F. Bettenstaedt et al., 1962: 359, pl. 5, fig. 5
		Miliammina sp.	A. R. Loeblich & H. Tappan, 1964: 220—221, fig. 134, 1—2
	calcareous, perforate	*Lenticulina* sp.	A. R. Loeblich & H. Tappan, 1964: 518—520, fig. 406, 6—8
		Ammonia becarii (Linné, 1758)*'	D. Schnitker, 1974: 216—223, pl. 1, fig. 1—9
		Elphidium crispum (Linné, 1758)*	J. Hageman, 1979: 94, pl. 5, fig. 6a—b
		Elphidium antoninum (D'Orbigny, 1846)*	A. Papp, 1963: 262, p. 10, fig. 3—5
		Elphidium articulatum (D'Orbigny, 1846)*	H. Hageman, 1979: 94, p. 5, fig. 4a—b
		Elphidium minutum (Reuss, 1865)*	J. Hageman, 1979: 96, p. 6, fig. 10a—b
		Elphidium cf. *pseudolessonii* Dam & Reinhold, 1941*	A. ten Dam & T. Reinhold, 1941a: 53, pl. 3, fig. 10a—b, pl. 6, fig. 11
		Elphidium cf. *rugosum* (D'Orbigny, 1846)*	A. Tollmann, 1957: 186, pl. 2, fig. 2
		Eponides umbonatus (Reuss, 1851)	J. A. Cushman, 1931: 52, pl. 11, fig. 1—3
		Cibicides boueanus (D'Orbigny, 1846)	J. H. van Voorthuysen & K. Toering, 1969: 100, pl. 3, fig. 1
		Nonion scaphum (Fichtel & Moll, 1798)	A. Tollmann, 1957: 183—184
		Nonion cf. *granulosum* (Dam & Reinhold, 1941)	A. ten Dam & T. Reinhold, 1941b: 211, fig. 1
		Nonion sp.	H. J. Hansen & F. Rögl, 1980: 173—174, pl. 1, fig. 1—19

The foram population of the autochthonous OMM silts is dominated by those species of Ammonia, Elphidium, and Nonion (Fig. 11) that are indicative of brackish water (Pokorny, 1958). Both, the foram assemblage and the ostracode fauna e. g., Cyprideis, Loxoconcha (Oertli, 1963; Hartmann, 1963) stand for a miohaline to pliohaline environment for the autochthonous Tertiary. This corresponds to the observed facies change from marine to limnetic. In contrast, the foram population in the displaced OMM silt clasts reveals a greater diversity. While 4 species of Elphidium and Nonion were found in the autochthonous OMM silts, 5 additional species of these genera were observed in the OMM clasts. The additional foram species in the OMM silt clasts, such as Elphidium minutum REUSS (Hageman, 1979), lived in an environment with increased salinity (brachyhaline). The foram fauna of OMM sand clasts is similar to that of the OMM silts (Table V). Thus all OMM clasts in the Bunte Breccia of drill core D are characterized by foram assemblages indicative of a brachyhaline facies while the autochthonous Tertiary formed in a mio- to pliohaline environment. The OSM sands, silts, clays, and limestones in the Bunte Breccia and in the autochthonous Tertiary display fossils indicative of a limnetic environment. Microfossils, however, are extremely rare. The OSM sands contain abundant

Table Va

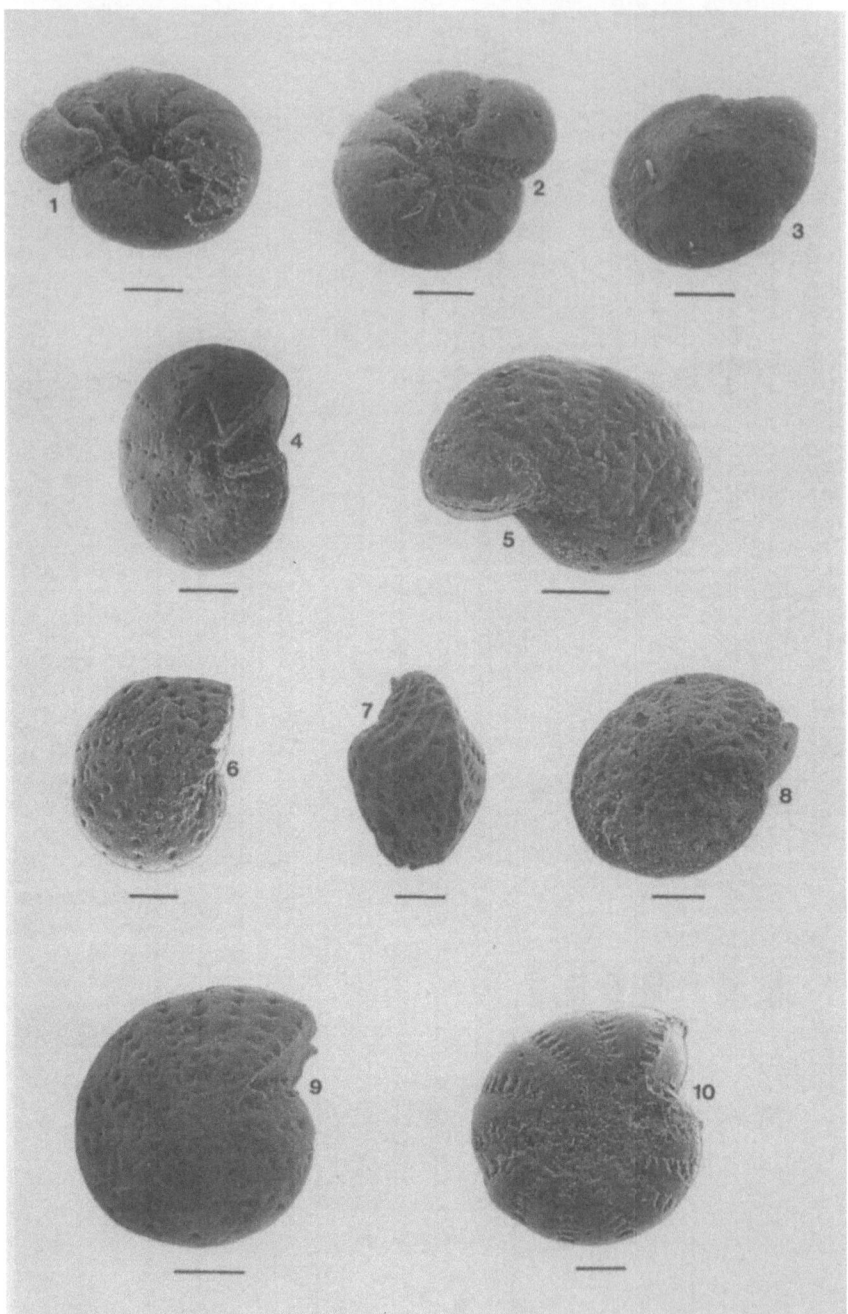

Fig. 11

Foraminifera in pure OMM clasts and autochthonous OMM of drill core D, SEM photographs; scale bar = 100 microns; 1) Ammonia beccarii (Linné, 1758), ventral view, 36.65 m; 2) Ammonia beccarii (Linné, 1758), ventral view, 36.50 m; 3) Ammonia beccarii (Linné, 1758), dorsal view, 37.00 m; 4) Elphidium minutum (Reuss, 1865), 25.50 m; 5) Elphidium cf. pseudolessonii Dam & Reinhold, 1941 a, 30.00 m; 6) Elphidium cf. rugosum (D'Orbigny, 1846), 4.50 m; 7) Elphidium crispum (Linné, 1758), marginal view, 20–21 m; 8) Elphidium crispum (Linné, 1758), 58.50 m; 9) Elphidium articulatum (D'Orbigny, 1839), 5–6 m; 10) Elphidium antoninum (D' Orbigny, 1846), 31.15 m.

gastropode shell detritus and one specimen of Tectochara ulmensis STRAUB, a fresh-water alga. Three specimens of Cepaea sp. (gastropoda) were observed in the autochthon-ous OSM limestone close to the contact with the overlaying Bunte Breccia. A few weather-ed specimens of Ammonia beccarii (LINNÉ) found in the freshwater clays at 16 m are remnants of reworked marine sediments (Table V).

The displaced OMM clasts evidently formed under environmental conditions that differed subtly from those of the autochthonous OMM bedrock. These results of the paleonto-logical studies imply a horizontal as well as a vertical transport of the displaced clasts. This holds true also for the OSM sands which were only found as clasts in the Bunte Breccia and were not encountered in the autochthonous bedrock of drill core D. There-fore not only the crater derived clasts but also large locally derved clasts have been trans-ported horizontally during deposition of the Bunte Breccia.

Broken foram shells and lignite pieces in the fine-grained groundmass of the polymict breccia (grain size < 2 mm) support the results of grain size, heavy mineral, and carbonate content analyses which indicated local clastic sediments as source lithologies of this groundmass.

Mixing Calculations and the Amount of Locally Derived Material

Based on the heavy mineral content and the garnet/epidote ratio of the polymict breccia groundmass and the Tertiary clastic sediments, least square mixing calculations were per-formed to model the composition of the polymict breccia groundmass. It was assumed that only Tertiary clastic sediments contributed to the formation of the breccia matrix. The garnet/epidote ratio was taken as 0.35 for OMM sand and silt, and as 16 for the OSM sand. The total heavy mineral content of the grain size fraction 0.063–0.125 mm was taken as 3 wt.%, 1 wt.%, and 13 wt.%, respectively. The results of these mixing calcu-lations are listed in Table VI for 11 polymict breccia sections. The average values of 72 wt.% OMM silt, 11 wt.% OMM sand, and 17 wt.% OSM sand are consistent with grain size data and the carbonate content for the polymict breccia groundmass. No additional components are required to reproduce the observed grain size fraction < 2 mm of the

Table VI: Mixing rations of OSM sand: OMM sand: OMM silt in the Bunte Breccia groundmass as determined by least-square mixing calculations on the basis of the heavy mineral content.

sample (depth, m)	OSM sand (%)	OMM sand (%)	OMM silt (%)
6.75	8.52+/−0	23.61+/− 1.38	67.87+/−0.87
12.15	17.75+/−1.06	16.34+/− 0	65.91+/−0.60
16.95	17.67+/−1.21	15.49+/− 0	66.85+/−1.44
18.40	14.41+/−0	36.59+/− 6.93	49.00+/−1.13
22.00	9.51+/−0	63.89+/−15.88	26.60+/−1.83
22.50	4.67+/−0.01	− 5.37+/−13.39	100.71+/−0.10
25.95	19.86+/−0	30.66+/− 0.17	49.48+/−0.07
34.95	22.62+/−0	5.09+/− 2.74	72.30+/−1.78
38.30	21.46+/−0	21.49+/− 0.13	57.05+/−0.15
48.80	58.34+/−1.54	25.13+/− 0	16.52+/−1.87
52.00	27.69+/−0	8.84+/‾ 0	63.47+/−0
average:	16.61+/−0	11.26+/− 2.34	72.13+/−0.59

Bunte Breccia. Linear analyses of thin sections of the polymict breccia groundmass yielded 4 to 14 vol. % clay and limestone clasts in the grain size fraction 1-2 mm. To account for the fact that these clasts could be crater derived, the total amount of locally derived material in the <2 mm grain size fraction of the Bunte Breccia is very conservatively taken as 90 vol. %.

The total amount of locally derived material in drill core D was calculated on the basis of the following observations and assumptions:

* 90 vol. % of the fine-grained polymict breccia consist of a groundmass with a grain size <2 mm
* 90 vol. % of this groundmass is locally derived
* the mean size of clasts 2-28 mm is 16 mm
* the mean size of clasts 28-200 mm is 60 mm
* all Tertiary material is of local origin
* all Mesozoic and older rocks are crater derived.

Based on these data and assumptions, 22 vol. % of drill core D consist of crater derived material. Allowing an uncertainty of 10 %, a total of 70—90 vol. % of the Bunte Breccia deposits at a distance of 2 crater radii from the center of the Ries crater are locally derived. This result is in agreement with the data obtained by Hörz et al., (1983) for this and other drill cores of the Ries drilling program.

5. Conclusions

Composition of the Bunte Breccia

The lithologic composition of the Bunte Breccia at a distance of two carter radii from the crater center comprises all stratigraphic units apparent in the target area down to a depth of more than 600 m. The major part of the continous deposits, however, consists of locally derived material. Lithic fragments of the crystalline basement and the lowermost sedimentary strata are rare. This is in accordance with data from experimentally produced craters (Stöffler et al., 1975) which also reveal decreasing amounts of deepseated material in the ejecta. The scarcity of crystalline basement rocks, however, may also be due to the strength discontinuity between sedimentary rocks and the crystalline basement rocks at a depth of about 600 m. This discontinuity may have triggered the formation of a terraced crater which was wider in the sedimentary strata than in the crystalline basement. The crystalline rocks would then have been ejected in a steeper angle than the sediments and thus were not transported as far (Hörz and Ostertag, 1979). The high total of 70—90 % of locally derived materials might to some extent be an effect of the unconsolidated nature and the thickness (tens of meters) of the Tertiary sediments south of the crater. At drill sites to the East of the crater such as at Otting and Itzing (Chao et al., 1977) the amount of locally derived material appears to be less. This may be explained by the fact that in this area unconsolidated Tertiary capped Upper Jurassic limestone only as a thin veneer. In addition, the underlying competent Upper Jurassic limestone was certainly not easily eroded and comminuted to produce a fine-grained locally derived breccia groundmass. The results obtained from the analyses of the Bunte Breccia drill cores taken south of the cliff line are highly significant for the interpretation of lunar breccias in general and the amount of locally derived materials contained therein, because the lunar regolith is unconsolidated. Therefore locally derived material is expected to become incorporated into lunar fragmental breccias in an amount very much like the Tertiary clastic sediments have been incorporated into the Bunte Breccia. Generally speaking, although the amount of locally derived material in the continous deposits of large impact craters may vary due to different material properties of the local country rock, the result that local material is a major if not the dominant component of the outer impact formations is of high significance for the interpretation of extraterrestrial samples as well as for photogeologic and remote sensing studies.

Breccia Deposition

The ejecta emplacement is a highly turbulent process. There is no systematic trend in the deposition of crater derived breccia components neither in size, orientation, nor in stratigraphic position. An inversion of the primary stratigraphic sequence of crater derived components is observed in experimentally produced craters (Stöffler et al., 1975), on the rim of small craters (Roddy et al., 1975) or in the inner ring of the Ries crater (Pohl et al., 1977), but it does not exist in the Bunte Breccia continuous deposits outside the actual crater rim. The thorough mixing of locally derived material to produce the breccia groundmass also required a highly turbulent breccia emplacement mechanism. The clasts must have suffered vastly different stress levels. Undeformed as well as twisted or elongated clay clasts of the same stratigraphic unit may be observed in the same poly-mict breccia section, and the Tertiary sand which was thoroughly mixed with other clastic sediments to produce the breccia groundmass is also found as a huge clast in the drill core. The local country rock was stripped during breccia transport down to a depth of several meters and this material was incorporated in the breccia. There was a considerable vertical and horizontal transport of locally derived material. Vertical transport of locally derived material by tens of meters is evident from the OMM and OSM clasts in the drill core which even display an inversion of their primary stratigraphic position. Horizontal transport of locally derived clasts is proven by limestone pebbles from the cliff line which are now deposited 6-7 km further to the South, and by the results of paleontological studies of locally derived clasts. Hörz et al. (1983) state that mixing of local material must have occurred on several different places involving slightly different locally derived source materials to form the four different types of fine-grained groundmass. They therefore stress the fact that the Bunte Breccia does not consist of clasts and a common matrix, but rather represents a deposit of lithic clasts and several kinds of breccia clasts. The breccia emplacement mechanism also involved the formation of "breccia within breccia" textures. It is very important to note that these textures formed at only one single impact event. Less than 1 % of the components of the Bunte Breccia do show evidence of shock at all. The peak shock pressures were 40–50 GPa at most (shock stage II, Stöffler, 1971). The post-shock temperature for feldspars shocked to this peak pressure is only of the order of $500°C$ (Ahrens et al., 1969). This low temperature combined with the overall low abundance of shocked material clearly indicates a deposition of the continuous deposits of the Ries crater at ambient temperatures. Detailed investigations of the crystalline basement rocks of all drill cores (Hörz and Banholzer, 1980) support the conclusion that 90 % of all impact-displaced masses during the Ries event have been deposited cold. The decision as to which transport mechanism was effective to produce the Bunte Breccia continuous deposits may be made by checking the amount of locally derived Tertiary material in these deposits. The high amount of locally derived material in the Bunte Breccia which even increases downrange (Hörz et al., 1983) was predicted by the secondary cratering hypothesis (Oberbeck, 1975) and is not consistent with a roll-and-glide mode of transport which was proposed by Chao (1976).

Acknowledgements

The National Aeronautics and Space Administration (NASA) provided the drill core. Helpful discussions with H. Gall (München), as well as with F. Hörz (Houston), and M. Kaever and D. Stöffler (Münster) are gratefully acknowledged. Part of the work was financially supported by the German Science Foundation (DFG). This paper is Research Group "Earth-Moon-System" Contribution No. 69.

References

Bettenstaedt, F., H. Fahrion, H. Hiltermann, and W. Wick, 1962: Tertiär Norddeutschlands (Abschn. B 9), in: Leitfossilien der Mikropaläontologie, Berlin, 339—378.

Birzer, F., 1969: Molasse und Ries-Schutt im westlichen Teil der Südlichen Frankenalb. Geol. Bl. NO-Bayern 19, 1-28.

Bolten, R. and D. Müller, 1969: Das Tertiär im Nördlinger Ries und in seiner Umgebung. Geologica Bavarica 61, 87—130.

Brestenska, E., 1974: Die Foraminiferen des Sarmatien s. str., Chronostratigraphie und Neostratotypen, Miozän M 5, IV. Bratislava, 243—270.

Chao, E.C.T., 1976: Mineral-Produced High-Pressure Striae and Clay Polish: Key Evidence for Non-ballistic Transport of Ejecta from Ries Crater. Science 194, 615—618.

Chao, E.C.T., 1977: The Ries Crater of Southern Germany. A Model for Large Basins on Planetary Surfaces. Geol. Jb. A 43, 81 p.

Chao, E.C.T., R. Hüttner, and H. Schmidt-Kaler, 1977: Vertical Section of Ries Sedimentary Ejecta Blanket as Revealed by 1976 Drill Cores form Otting and Itzing (abstract). Lunar Science VIII, Lunar and Planetary Institute, Houston, 163—165.

Chao, E.C.T., R. Hüttner, and H. Schmidt-Kaler, 1978: Principal Exposures of the Ries Meteorite Crater in Southern Germany. Description, photographic documentation and interpretation. Bayer. Geol. Landesamt, München.

Cicha, I., I. Zapletalova, A. Papp, J. Ctyroka, and R. Lehotayova, 1971: Die Foraminiferen der Eggenburger Schichtengruppe (incl. Arcellinida), Chronostratigraphie und Neostratotypen, Miozän M 1, II. Bratislava, 234—355.

Cicha, I., F. Rögl, J. Ctyroka, I. Zapletalova, and A. Papp, 1973: Die Foraminiferen des Ottnangien, Chronostratigraphie und Neostratotypen, Miozän M 2, III. Bratislava, 297—355.

Cushman, J.A., 1931: The Foraminifera of the Atlantic Ocean: Part 8. Bull. U.S. Natl. Mus. 104, 1—179.

Dam, A. ten and T. Reinhold, 1941a: Die stratigraphische Gliederung des niederländischen Pliozäns/Pleistozäns nach Foraminiferen. Meded. geol. Sticht., Ser. C-V, 1, 66 p.

Dam, A. ten and T. Reinhold, 1941b: Nonionidea as Tertiary index-foraminifera. Geol. en Mijnb., N.S. 3, 6, 209—212.

Folk, R.L. and W.C. Ward, 1957: Brazos River Bar: A Study in the significance of Grain Size Parameters. J. Sediment. Petrol. 24, 3—26.

Gall, H., 1971: Geologische Karte von Bayern 1:25 000, Erläuterungen zum Blatt Nr. 7328 Wittislingen. Bayerisches Geologisches Landesamt, München.

Gall, H., 1974: Neue Daten zum Verlauf der Klifflinie der Oberen Meeresmolasse (Helvet) im südlichen Vorries. Mitt. Bayer. Staatssamml. Paläont. hist. Geologie 14, 81-101.

Gasse, W., 1981: Die Tertiären Foraminiferen der NASA-Forschungsbohrung Demmingen (SW-Vorries). Diploma thesis, University of Münster, Münster.

Graup, G., 1978: Das Kristallin im Nördlinger Ries, Petrographische Zusammensetzung und Auswurfmechanismus der kristallinen Trümmermassen, Struktur des kristallinen Untergrundes und Beziehungen zum Moldanubikum. Enke Verlag, Stuttgart, 190 p.

Hageman, J., 1979: Benthic foraminiferal assemblages from Plio-Pleistocene open bay to lagoonal sediments of the Western Peloponnesus (Greece). Utrecht micropaleont. Bull., 20, 171 p.

Hagn, H., 1981: Die Bayerischen Alpen und ihr Vorland in mikropalänontologischer Sicht, Geologica Bavarica 82, 408 p.

Hansen, H.J. and F. Roegl, 1980: What is Nonion? Problems involving foraminiferal genera described by Montfort, 1808 and the type species of Fichtel and Moll, 1798. J. Foram. Res. 10, 3, 173—179.

Hartmann, G., 1963: Zur Morphologie und Ökologie rezenter Ostracoden und deren Bedeutung bei der Unterscheidung mariner und nichtmariner Sedimente. Fortschr. Geol. Rheinl. Westf. 10, 67—80.

Kiderlen, H., 1931: Beiträge zur Stratigraphie und Paläogeographie des süddeutschen Tertiärs. N. Jb. Min. Geol. Paläont., Beil. Bd. 66, Abt. B, 215—384.

Hiltermann, H, 1949: Klassifikation der natürlichen Brackwässer. Erdöl und Kohle 2, 4—8.

Hörz, F. and G.S. Banholzer, 1980: Deep seated target materials in the continuous deposits of the Ries Crater, Germany. Proc. Conf. Lunar Highlands Crust (Papike, J.J. and R.B., Merrill, eds.). Pergamon Press, New York, 211—231.

Hörz, F. and R. Ostertag, 1979: The transient cavity of the Ries crater, Germany (abstract). Lunar Planet. Sci. X, Lunar and Planetary Institute, Houston, 570—572.

Hörz, F., H. Gall, R. Hüttner, and V.R. Oberbeck, 1977: Shallow Drilling in the "Bunte Breccia" Impact Deposits, Ries Crater, Germany. In: Impact and Explosion Cratering (Roddy, D.J., R.O. Pepin and R.B. Merrill, eds.). Pergamon Press, New York, 425—448.

Hörz, F., R. Ostertag, and D.A. Rainey, 1983: Bunte Breccia of the Ries: Continuous deposits of large impact craters. Rev. Geophys. Space Phys. 21, 1667—1725.

Hüttner, R., 1961: Geologischer Bau und Landschaftsgeschichte des östlichen Härtsfeldes (Schwäbische Alb). Jh. geol. Landesamt Baden-Württemberg, 4, 49—125.

Köhler, H., G. Graup, P. Horn, and D. Müller-Sohnius, 1981: Rb-Sr-Mineralalter aus der Forschungsbohrung Nördlingen 1973 (abstract). Fortschr. Miner. 59, 93—94.

Lemcke, K., W. v. Engelhardt, and H. Füchtbauer, 1953: Geologische und sedimentpetrographische Untersuchungen im Westteil der ungefalteten Molasse des süddeutschen Alpenvorlandes. Beih. Geol. Jb. 11.

Loeblich, A.R. and H. Tappan, 1964: Treatise on Invertebrate Paleontology C 2. Lawrence, Kansas, 2, 900 p.

Morrison, R.H. and V.R. Oberbeck, 1978: A Composition and Thickness Model for Lunar Impact Crater and Basin Deposits. Proc. Lunar Sci. Conf. 9th, 3763—3785.

Oberbeck, V.R., 1975. The Role of Ballistic Erosion and Sedimentation in Lunar Stratigraphy. Rev. Geophys. and Space Phys. 13, 337—362.

Oertli, J., 1963: Fossile Ostracoden als Milieuindikatoren. Fortschr. Geol. Rheinl. Westf. 10, 53—66.

Papp, A., 1963: Die biostratigraphische Gliederung des Neogens im Wiener Becken. Mitt. geol. Ges. Wien 56, 225—317.

Pohl, J., D. Stöffler, H. Gall, and K. Ernstson, 1977: The Ries Impact crater. In: Impact and Explosion Cratering (Roddy, D.J., R.O. Pepin, and R.B. Merrill, eds.). Pergamon Press, New York, 343—404.

Pokorny, V., 1958: Grundzüge der zoologischen Mikropaläontologie 1 + 2. Berlin.

Roddy, D.J., J.M. Boyce, G.W. Colton, and A.L. Dial, 1975: Meteor Crater, Arizona, Rim Drilling with Thickness, Structural Uplift, Diameter, Depth, Volume, and Mass Balance Calculations. Proc. Lunar Sci. Conf. 6th, 2621—2644.

Salger, M. and H. Schmidt-Kaler, 1973: Sedimentologische Untersuchungen im Lias von Altdorf bei Nürnberg (fränkische Alb). Geologica Bavarica 67, 162—168.

Salger, M. and H. Schmidt-Kaler, 1974: Sedimentologische Untersuchung im unteren Lias anhand der Kernbohrung Ettenstatt nordöstlich von Weissenburg in Bayern. Geol. Bl. NO-Bayern 24, 191—194.

Schetelig, K., 1962: Geologische Untersuchungen im Ries. Das Gebiet der Blätter Donauwörth und Genderkingen. Geologica Bavarica 47, 98 p.

Schmeer, D., 1955: Sedimentpetrographische Beobachtungen aus der Oberen Süßwassermolasse im Bereich von Freising bis Landshut. Z. deutsch. Geol. Ges. 105, (1953), 496—516.

Schmidt-Kaler, H., 1969a: Versuch einer Profildarstellung für das Rieszentrum vor der Kraterbildung. Geologica Bavarica 61, 38—40.

Schmidt-Kaler, H., 1969b: Keuper und Jura in der Tiefbohrung Riedenburg. Geol. Bl. NO-Bayern 19, 97—112.

Schneider, W., 1971: Petrologische Untersuchungen der Bunten Breccie im Nördlinger Ries. N. Jb. Miner. Abh. 114, 136—180.

Schnitker, D., 1974: Ecotypic variation in Ammonia becarii (Linné). J. foram. Ges. 4, 217—223.

Schnitzer, W.A., 1953: Sedimentpetrographische Untersuchungen an den postjurassischen Überdeckungsbildungen der mittleren südlichen Frankenalb. Geol. Bl. NO-Bayern 3, 121—134.

Schröder, B., 1962: Schwermineralführung und Paläogeographie des Doggersandsteins in Nordbayern. Erlanger geol. Abh. 42, 3—29.

Steininger, F., F. Rögl, and E. Martini, 1976: Current Oligocene/Miocene biostratigraphic concept of the Central Parathetys (Middle Europe). Newslett. Stratigr., Berlin 4, 174—202.

Stöffler, D., 1971: Progressive metamorphism and classification of shocked and brecciated crystalline rocks at impact craters. J. Geophys. Res. 76, 5541—5551.

Stöffler, D., D.E. Gault, J. Wedekind, and G. Polkowski, 1975: Experimental hypervelocity impact into quartz sand: distribution and shock metamorphism of ejecta. J. Geophys. Res. 80, 4062—4077.

Tollmann, A., 1957: Die Mikrofauna des Burdigal von Eggenburg (Niederösterreich). Sitzungsber. österr. Akad. Wiss., math.-natw. Kl., Wien, 166, 165—213.

Woorthuysen, J.H. van and K. Toering, 1969: Distribution quantitative des foraminifères neogènes et quarternaires aux environs d'Anvers. Meded. Rijks geol. Dienst, N.S., Haarlem, 20, 93—123.

The Rb-Sr-Age of the Rochechouart Impact Structure, France, and Geochemical Constraints on Impact Melt-Target Rock-Meteorite Compositions

Wolf Uwe Reimold*

WITS-CSIR Schonland Research Centre of Nuclear Sciences, University of the Witwatersrand,
1 Jan Smuts Avenue, Johannesburg 2000, South Africa.

Wolfgang Oskierski

Institute of Geology, University Münster, Corrensstr. 24, D-4400 Münster, FRG.

Key Words

Rochechouart crater
Impact crater
Impact melt
Impact breccia dike
Chronology
Rb-Sr method in geochronology
Chondrite
Mixing calculation
Cratering flux
Alteration

Received: February 1985, accepted: April 1985.

Abstract

The *age* of the Rochechouart impact event was determined to be 185.5 ± 4.4 (1 σ) Ma by the Rb-Sr method applied to an impact melt rock sample of the coherent melt body from the center of the impact structure. Despite the fact that the analytical data define a mixing line between actual impact melt and alteration product compositions, we believe that this age is of geochronological significance.

The determination of a mafic component contributing to the *target rock mixture* from which the impact melt was formed yielded new evidence for a chondritic nature of the meteoritic projectile.

Mixing calculations for *impact melt and clastic polymict impact breccia dikes* from the crystalline basement of the Rochechouart crater structure revealed the origin of the dike fillings and gave clues on the time and formation processes of dikes within the cratering cycle. A SEM and X-ray study of the coarse-grained impact melt sample provided further insight into crystallization processes and, together with geochemical calculations, show that the average impact melt composition was determined by crystallization differentiation rather than by weathering processes as previously assumed.

Introduction

The geological structure of Rochechouart in the Haut-Limousin/NW Massif Central, France (Fig. 1) was first considered as being of possible meteorite impact origin by Kraut (e. g. 1969). Since then the impact origin of this approximately 20–25 km wide, deeply eroded crater structure has been confirmed beyond doubt by the petrographic and geochemical studies of Lambert (1974, 1977) and by the identification of a probably chondritic composition for the projectile by Palme et al. (1980). A re-investigation of the distribution of impact breccias within the crater interior, and of impact breccia dikes within the crystalline basement, was recently completed by Oskierski (1983) and Oskierski and Bischoff (1983). The original crater morphology has been destroyed by progressive erosion. Mainly to the West and North of the probable point of impact subhorizontal deposits of impact formations, namely clastic monomict and clastic polymict breccias, suevite and impact melt, can be mapped (Lambert, 1974, 1977). The maximum thickness of these more or less coherent nappes can be up to 60 m. Impact melt is found mainly in the central crater region. Two varieties of impact melt, clast-poor and clast-rich, have been described (Oskierski, 1983). Both impact melt types display maximum thicknesses of only a few meters. and the melt blankets were deposited either in contact with the crystalline basement or on top of clastic polymict breccias.

Fig. 1
Geographical location of the Rochechouart impact structure (R) in the NW part of the French Massif Central, called Haut-Limousin.

The crystalline basement which is often clearly exposed is composed mainly of low- to medium-grade metamorphic gneisses of granitic to granodioritic composition, of leptynites, and of small lenticular complexes of diorite and amphibolite. In addition, a postvariscan granitization phase supplied microgranite and pegmatitic granitic dikes. However, despite the extensive work of Lambert (1974, 1977) and Bischoff and Oskierski (this volume), a number of unsolved problems remained.

(1) What is the age of this impact structure? According to K-Ar and fission track dating cited by Lambert (1977) it is likely that the event occured during the Jurassic-Triassic period, most probably within the time span from less than 155 to 260 Ma (Lambert, 1982). The K-Ar dating results of Lambert (1977) did not yield an undisputable age datum, because of the severe alteration of the analysed impact melt samples and because impact breccias as analysed by Lambert most probably were not completely reset during the cratering event (e. g. Jessberger et al., 1978; Bogard et al., 1981).

(2) Is the nature of the meteorite projectile chondritic? Janssens et al. (1977) favored an Iron IIA meteorite projectile, whereas the evidence of Palme et al. (1980) and of Horn and El Goresy (1980) requires a chondritic projectile. In order to determine the meteoritic contamination of the impact melt in the center of a meteorite crater, Palme et al. (1979) and Reimold (1982) have shown that detectable enrichments of siderophile elements — a meteorite signature — can only be expected in coherent impact melt sheets of crater centers. It is necessary to determine precisely the indigenous contribution (Palme, 1980), i. e. the abundance of siderophile elements such as Ni, Co, Cr, Ir, or Au — characteristic for meteorite compositions — in the molten target rock mixture. Previous calculations of the Rochechouart indigenous component (Palme et al., 1980) were based on the assumption of Lambert (1977, 1982) (derived from geological observations, counting statistics of target rock clasts in impact breccias, and from two-component (gneiss, granite) chemical mixing calculations), that the probable target was composed only of gneiss and granite with proportions between 50—90 % and 5—50 %, respectively. However, Lambert (1977) also describes small outcrops of mafic rocks such as diorites and amphibolites within the crater region, and mafic rocks are likely to carry considerable amounts of siderophile elements.

(3) The occurrence of impact breccia dikes within the crystalline basements of terrestrial impact structures is a widespread phenomenon (e. g. Dence, 1965; Rehfeldt, 1983; Stöffler, 1977; Wilshire et al., 1972). The Rochechouart breccia dikes have been classified structurally and petrographically by Oskierski (1983). However, it was still ambiguous *how to interpret the derivation of the fillings* (from adjacent host rock, injected impact melt, locally produced melt, mixtures thereof?), and at *what time during the cratering process dike formation took place*. The importance of these questions is evident in view of the fact that dimict breccias interpreted as impact breccia dikes are an important constituent of the lunar highland sample collection and form chemically distinct melt groups therein (e. g. McKinley et al. 1981, 1982, 1983).

(4) The average chemical and petrographic composition of the impact melt from Rochechouart is puzzling to the extent that the major element chemistry — especially with respect to alkali element concentrations — cannot be explained by any mixing models involving the outcropping basement rocks (Lambert, 1982). It has therefore been assumed that observed compositions were severely influenced by weathering and hydrothermal alteration processes.

In order to contribute to the solution of these problems, we searched for an impact melt rock suitable for petrographic and geochemical/geochronological analysis. In addition, the small amphibolite complexes were studied by petrographic and chronological methods (Reimold et al., in prep.) to provide data for a multi-component mixing model including *all* target rock varieties. Thus it was possible to determine the essential indigenous contribution of siderophile elements. Further mixing calculations were performed to reveal the component proportions contributing to dike breccia fillings. Finally, bearing all analytical results in mind, it was possible to address objektive (4): the chemical evolution and modification of the Rochechouart impact melt.

Description of sample 5/19 and experimental procedures

The coarse-grained (up to 450 μm) impact melt rock 5/19 from Babaudu ($0°47'30''$W/ $45°49'20''$N) close to the center of the structure is characterized by a subophitic matrix texture prominantly formed by feldspar laths (Fig. 2) intergrown with low Ca-pyroxene. Interstices are filled with more or less altered mesostasis. Alteration products were shown to be of saponite or montmorillonite composition. Small amounts of ilmenite and FeNi-sulfide grains complete the mode of the matrix. The rare clasts consist mainly of fractured or recrystallized quartz with minor amounts of plagioclase or K-feldspar (less than 2 vol% of total fragment population). A modal analysis is presented in Table 1. Mineral identi-

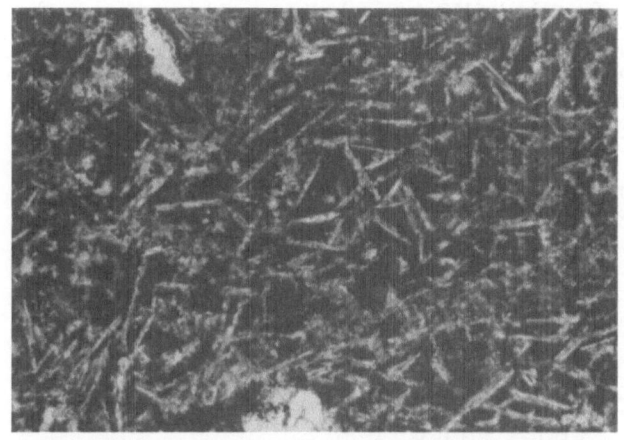

a)

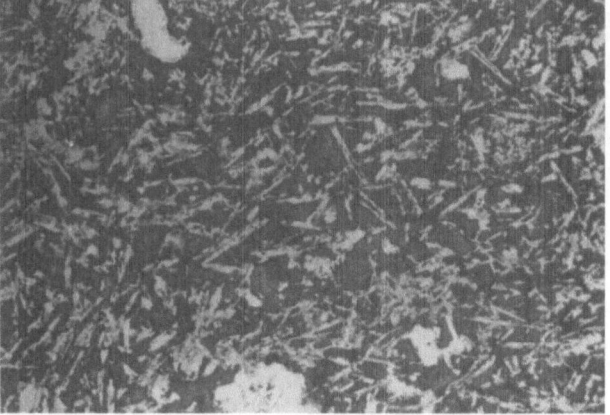

Fig. 2
Typical subophitic texture of Rochechouart impact melt rock 5/19 with lathy sanidine (white), prismatic-platy pyroxene (light-gray), and altered interstitial mesostasis (dark). Upper left: a vesicle rimmed by oxides and clay minerals; at lower margin: a partially recrystallized quartz clast; width of field of view: 1100 μm;
a) plane polarized light,
b) reflected light.

b)

Table I: The mineralogical composition of impact melt rock 5/19
as determined by pointcounting

Mineral	vol.%
feldspar	42.0
Ca-pyroxene	20.0
mesostasis	20.0
ilmenite	1.0
FeNi-sulfide	2.0
clasts	15.0
Sum	100.0

fication was supported by SEM-ED techniques and Guinier-Jagodzinski X-ray diffraction.
In comparison to typical impact melt rock samples from Rochechouart, specimen 5/19 is
the most coarse-grained, clast-poor, and least altered sample observed to date. Characteristically Rochechouart impact melt rocks are composed of a very fine-grained to cryptocrystalline matrix and considerable amounts of mineral and lithic clasts, and matrices are
generally altered to mixtures of clay minerals, zeolites, and carbonate minerals. In sample
5/19, alteration is restricted to interstitial mesostasis and only marginally affects pyroxene.
In order to obtain a sample residue free of alteration products for dating by the Rb-Sr
isotope method, a 50 g aliquot of crushed 5/19 material was treated according to the
diagram of Fig. 3. Feldspar and pyroxene fractions were enriched by magnetic and heavy
liquid separation techniques from grain size fraction 180–250 μm followed by hand-
picking. Bulk rock splits and mineral separates were leached in 0.5 N HCl, to remove clay
mineral alteration products, without significantly affecting feldspar and pyroxene (cf.
Figs. 6a-d).
Rb-Sr isotope analyses were performed at the Institute of Mineralogy in Münster following
the procedures of Reimold (1982) or Jarrar et al. (1983). The average of 7 analytical

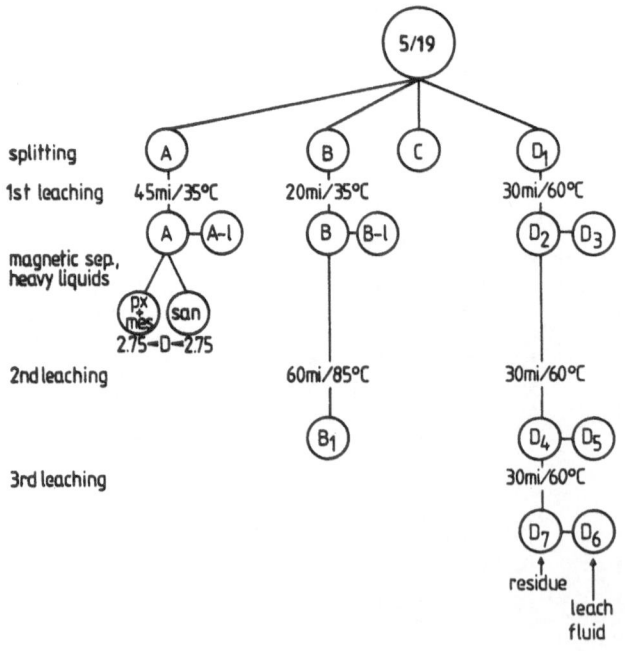

Fig. 3

Flow diagram illustrating
preparation (sample splitting and
mineral enrichment by magnetic
separation and heavy liquid treatment) and leaching (with 0.5 n HCl)
of impact melt rock 5/19 subsamples
for Rb-Sr isotope analysis.

Table II: Rb and Sr concentrations and isotopic compositions (of whole rock (WR) samples, mineral separates (A (D > 2.75) = pyroxene + mesostasis, A (D < 2.75) = sanidine), and leach fluids (LF) obtained from Rochechouart impact melt sample 5/19 according to procedures of Fig. 3. Concentration data given are raw analytical values that have only been normalized to weights of analysed samples given in column 1. Analysed samples were either aliquots of the total recovered residue fractions (Fig. 3) or the whole of a recovered leach solution.

Sample	Wt (mg)	Rb (ppm)	Sr (ppm)	$^{87}Rb/^{86}Sr$	$^{87}Sr/^{86}Sr$**	^{87}Sr (μmol/g)
5/19						
C (WR)	45.8	206.85± .71	39.77 ± .04	15.824 ± 71*	.75335±15*	.0321
B (WR)	37.3	255.36± .14	11.14 ± .01	67.484 ± 73	.89057±23	.011
B_1 (WR)	50.1	245.33±1.38	10.10 ± .003	70.915 ±265	.90070±15	.010
A (D<2.75)	3.8	258.27± .26	21.85 ± .01	34.517 ± 52	.80276± 3	.0196
A (D>2.75)	82.1	251.85± .6	10.39 ± .006	71.440 ± 21	.89697±12	.0103
A-l (LF)	89.2	11.11± .15	44.24 ± .01	.727 ± 10	.71286± 2	.0355
B-l (LF)	38.4	6.13± .003	4.165± .006	4.258 ± 6	.71676±22	.0034
D_1 (WR)	79.1	256.54± .32	33.73 ± .04	22.138 ± 53	.7694 ± 3	.029
D_2 (WR)	130.3	271.16± .62	14.66 ± .007	54.265 ±150	.85081±16	.014
D_3 (LF)	7.4	39.17± .09	1734.31 ± .3	.0654± 6	.71310± 9	1.39
D_4 (WR)	263.9	276.10± .36	14.145± .007	57.315 ±106	.8598 ± 5	.013
D_5 (LF)	24.8	21.15± .03	77.44 ± .03	.7905± 14	.71406±13	.062
D_6 (LF)	60.5	238.30± .81	19.080± .009	36.466 ±141	.80115± 7	.017
D_7 (WR)	219.6	263.38± .68	12.288± .005	63.042 ±187	.8772 ± 3	.013

* Errors are $2\sigma_m$ standard deviations and refer to last digits

** Normalized to $^{86}Sr/^{88}Sr = .1194$

values obtained for the NBS-SRM-987 Sr standard during the period of this study is 0.71037 ± 7 (1 σ), a value consistent with the average value obtained with the mass spectrometers of the laboratory since 1976 (e. g. Reimold et al., 1981). Total chemistry Sr blank values were of the order of 0.2—1.7 ng, and Rb blanks, of the order of 0.1—1.6 ng. The results are compiled in Table II.

For SEM characterization of sample 5/19 and its mineral separates, a Phillips SEM 505 electron microscope equipped with a semiquantitative energy-dispersive analytical unit was used at the Institute of Medical Physics, Münster. The spot size was generally 50—100 μm, acceleration voltage was 30 kV, and the effective take-off angles varied from 15—20°. ED counting times were integral 60 sec and average registered numbers of counts/sec were 1 500.

Multi-component mixing calculations were made by the HMX mixing program (Stöckelmann and Reimold, 1984). In contrast to previously used mixing programs, based on linear least-squares regression methods, the HMX procedure utilizes the harmonic least-squares theory (Kroll and Stöckelmann, 1979) which allows for analytical uncertainties of all input parameters, i. e. compositional data for components *and* mixture. Input data for target rocks and impact melts were taken from Lambert (1977, 1982), Janssens et al., (1977), Palme et al., (1980), and for dike breccia calculations from Oskierski (1983) and Oskierski and Bischoff (1983). In addition, a XRF analysis was obtained in this study for a bulk sample of 5/19 and several amphibolite specimens from the Rochechouart area. The latter analytical data will be published separately (Reimold et al., in prep.). In order to obtain the average content of siderophile elements in the Haut-Limousin amphibolites, a mixture of aliquots from four amphibolite samples was analysed by H. Palme at the INAA laboratory of the Max-Planck-Institute for Chemistry in Mainz. The XRF analysis of Si, Ti, Al, Fe, Mn, Mg, Ca, and P contents of 5/19 and amphibolites was performed on a Siemens SRS-200 X-ray fluorescence spectrometer. The maximum relative analytical uncertainties were obtained from repetitive analyses of several USGS, QMC, CRPG, ANRT, and GSJ standards and are for the above list of elements of the order of 0.6, 3, 1, 1, 5, 7, 1.5, and 3 % respectively. K and Na abundances were determined by means of a Pye-Unicam atomic absorption spectrometer to accuracies of ± 5 %.

Results

I. The age of the Rochechouart impact structure

The analytical data for the Rb-Sr isotope analyses on 5/19 impact melt and mineral separates are presented in the isochron diagram of Fig. 4 and in Table II. All data including whole rock, mineral and leach fluid samples plot on or close to a regression line, the slope of which corresponds to an age of 185.5 ± 4.2 Ma (1 σ error), and the initial Sr isotope ratio is 0.7116 ∓ 9 (2 σ) (calculated using the decay constant of 0.0142 AE^{-1}, and for whole rock samples only). Including the data for the two mineral separates shifts the age within the error limits of the first calculation to a value of 184.5 ± 3.0 Ma (I_{Sr} = 0.7114 ∓ 8). A York calculation for leached samples yields T = 185.6 ± 8.2 Ma (I_{Sr} = 0.7118 ∓ 9). Using the lithological composition of the Rochechouart target area as presented below and values of I_{Sr} for the average target rocks (gneiss, granite, amphibolite and diorites; data

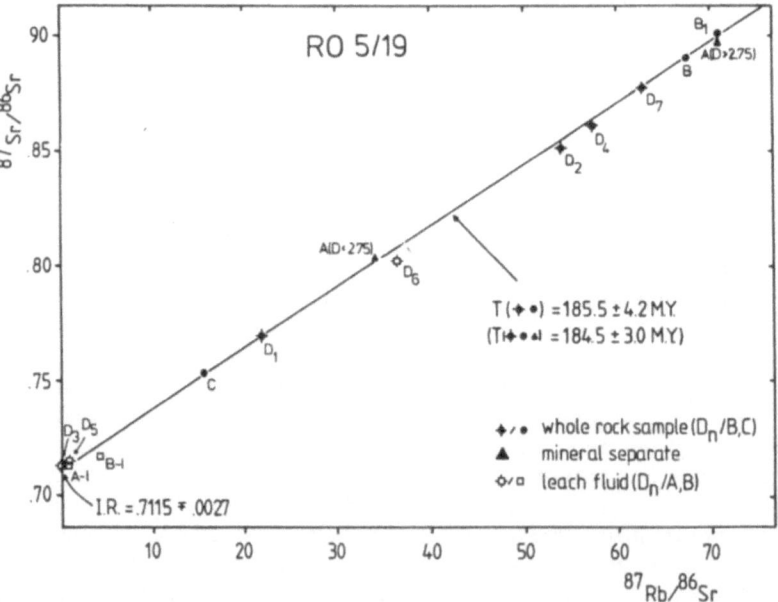

Fig. 4

Rb-Sr isotopic compositions of impact melt rock 5/19 (whole rock samples, mineral separates (A (D < 2.75 = sanidine Or$_{100}$; A (D > 2.75 = pyroxene ± mesostasis, cf. text), and leach fluids. Sample numbers correspond to those of Fig. 3, and symbols are used in following diagrams (Figs. 5a, c) with identical meanings. The isochron age of 185.5 ± 4.2 Ma was calculated for treated samples only using the decay constant λ = 0.0142 AE^{-1}; the I.R. was calculated for all data; cf. also text.

presented by Reimold et al. (1983) and from papers cited therein and recalculated to an age of 185 Ma), a Sr isotope ratio for the target rock mixture of 0.716–0.707 is calculated. The impact melt initial ratio of 0.712 lies well within this range.

However, the question arises whether this regression line represents a true isochron of real age significance or whether it is only a mixing line between original impact melt composition (D_7–B_1 leached whole rock samples, Fig. 4) and a young secondary alteration component possibly represented by the leach fluid data of low Sr isotopic compositions.

A first observation in Fig. 4 is that both data obtained from mineral separates plot on the regression line. As will be shown later, sample A (D < 2.75) is composed of pure K-feldspar, whereas sample A (D > 2.75) contains some mesostasis besides all the pyroxene. A small part of mesostasis in this separate is required to account for the high Rb/Sr ratio, as the pyroxene will be shown to be low-Ca pyroxene and cannot be expected to contain as much as 25 ppm Rb (Table II). As A (D > 2.75) plots remarkably close to late-leached whole rock sample B_1, thought to represent the original impact melt isotopic composition, one can be reasonably certain that no alteration product of mesostasis and, to a minor extent, of pyroxene was included in this separate. Thus the mineral phases of 5/19 seem to be well-equilibrated with respect to Sr isotopes, providing support for the interpretation of the 185 Ma age as the crystallisation age of the impact melt.

Sample 5/19 is composed of a holocrystalline matrix and 15 vol % target rock fragments that are mainly (> 98 %) composed of quartz, and it is unlikely that it contains significant amounts of radiogenic strontium inherited from precursor rocks. No geological activity in the Haut-Limousin after 250 Ma ago has been recorded that might have affected (partially reset?) the Sr isotope system of the Rochechouart target rocks (Reimold et al., 1983). The comparison of average impact melt (Oskierski and Bischoff, 1983) and sample 5/19 chemical compositions (Table III) shows that sample 5/19 is considerably less affected by alteration processes than the previously analysed samples, which is most obvious from the comparison of Al_2O_3, CaO and K_2O contents. A SEM study of treated and untreated 5/19 samples (Fig. 6) shows clearly that all alteration products were removed by leaching – prior to isotope analysis – of samples A, B, and B_1. The leaflike interstitial clay minerals of Figs. 6a and 6b cannot be seen in the SEM photographs of leached sample B_1 (Figs. 6c and 6d). Instead, shallow (mostly $1\,\mu m$) negative etching forms are visible on plagioclase and to a lesser extent on pyroxene surfaces, indicating that the major mineral constituents were also surficially dissolved in the leaching process. However, the rather low Rb and Sr concentrations of the leach fluids (Table II, Fig. 5) prove that the extent of dissolution was minimal.

In Fig. 5a the Rb and Sr concentrations of the 5/19 samples are presented. This presentation has been chosen, since a plot of the absolute concentrations does not enlighten the following point: obviously the leach fluids contain high and variable amounts of

Table III: Chemical composition of average impact melt from the Rochechouart area (Oskierski and Bischoff, 1983) and of sample 5/19.

	average impact melt (± 1σ)	5/19
SiO_2	66.6 ± 4.2	61.92
Al_2O_3	16.24 ± 2.4	16.89
Fe_2O_3	4.22 ± 2.2	4.60
MgO	1.29 ± 1.1	2.81
CaO	0.25 ± 0.2	0.37
Na_2O	0.34 ± 0.3	0.31
K_2O	10.19 ± 1.8	11.74
TiO_2	0.76 ± 0.1	0.80

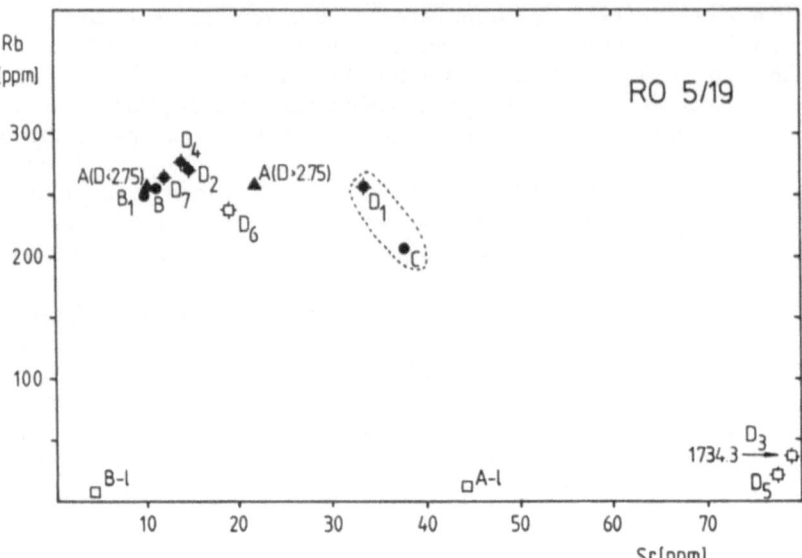

Fig. 5a

Rb and Sr concentrations as analyzed in 5/19 samples (symbols as explained in Fig. 4). This presentation was preferred to a diagram of absolute concentrations for scaling reasons only.

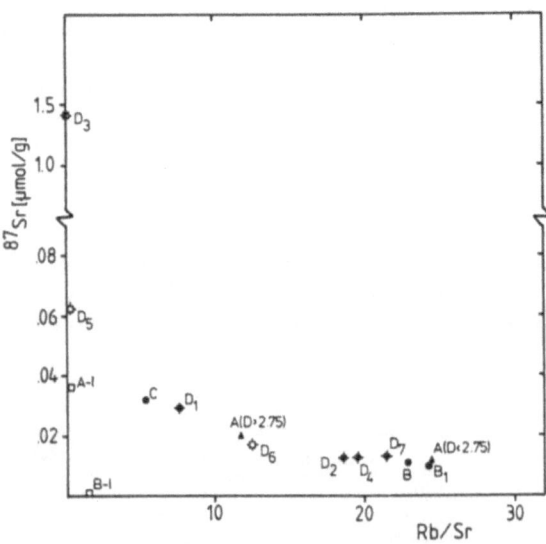

Fig. 5b

Correlation diagram of Rb/Sr ratios and [87]Sr concentrations for 5/19 samples (for explanation see text).

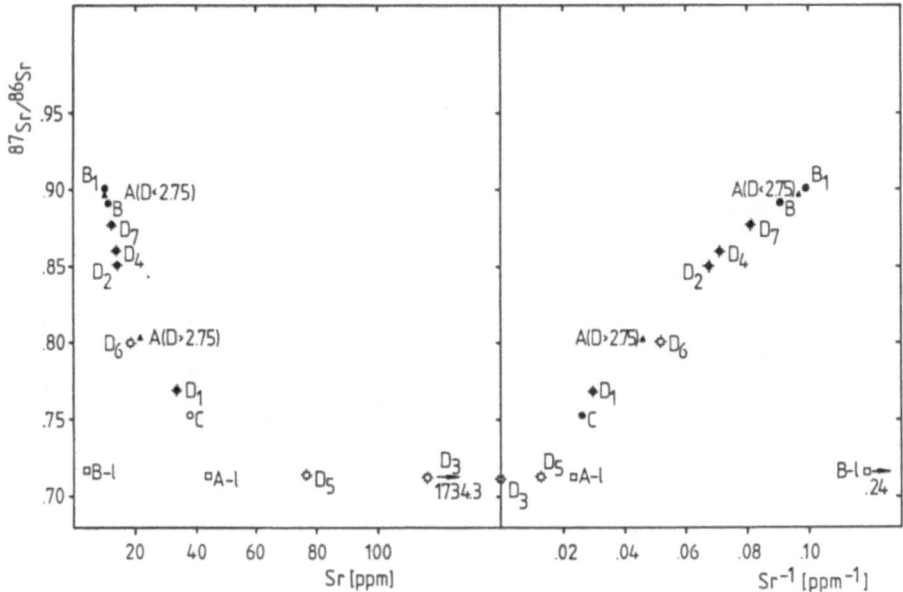

Fig. 5c

Mixing hyperbola ($^{87}Sr/^{86}Sr$ vs. (Sr)) and mixing line ($^{87}Sr/^{86}Sr$ vs. (Sr)$^{-1}$) for 5/19 samples according to Faure (1977).

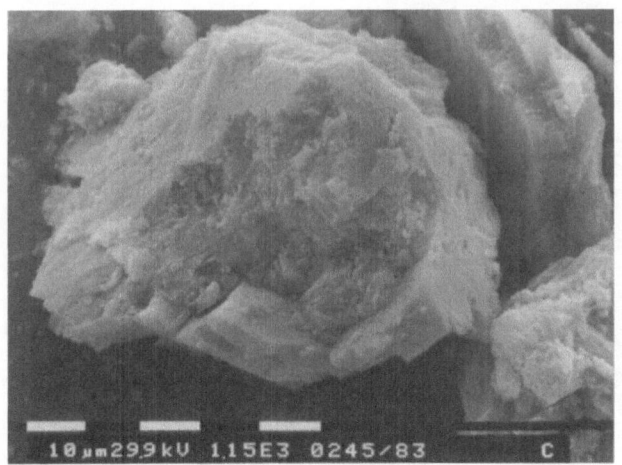

Fig. 6a

SEM photograph of untreated sample 5/19 grain. Patches of altered mesostasis are clearly visible in between crystalline plagioclase (left and right) and pyroxene (top of grain). Scale bar is equivalent to 10 μm.

Fig. 6b
Enlarged area of Fig. 6a showing leaf-like structure of alteration products (saponite/montmorillonite) and the slight surficial alteration of a pyroxene crystal (upper left). Scale bar = 10 μm.

Fig. 6c
A grain of strongly leached subsample B_1 (cf. Fig. 3). No clay mineral patches can be detected. Instead, negative etching forms are revealed on practically all crystal faces. Slae bar = 100 μm.

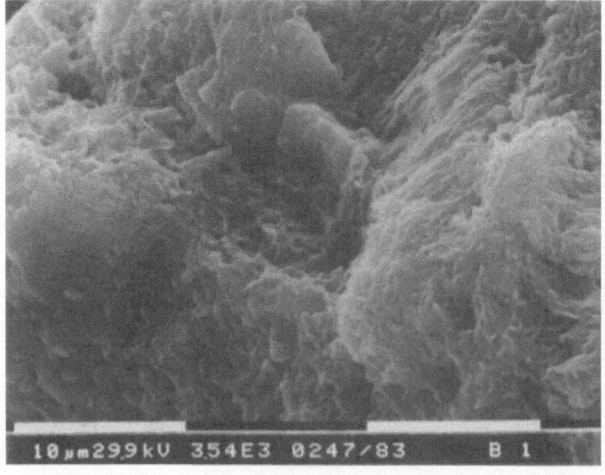

Fig. 6d
Relict grain after B_1 treatment (Fig. 3). Clayey mesostasis is totally removed (center) and clean sanidine (left and center top) and pyroxene (right) remain and display both concave etching forms. Scale bar = 10 μm.

Sr and rather low uniform concentrations of Rb. Sample B-*l* is unique in containing extraordinarily low Rb and Sr concentrations and also yields aberrant data points in the isotope diagram of Fig. 4, and in Figs. 5b and 5c; it is thought that Rb and Sr were partially lost from this sample during the chemical procedure. Sample D_7, plotting close to the leached impact melt samples (Fig. 4), most probably consists of nearly pure impact melt. In contrast, leach fluid D_3 with an extremely high Sr content most probably represents the alteration phase. Slight compositional differences are observed between unleached whole rock aliquots D and C despite the fact that the bulk sample was pulverized to 250 μm grain sizes. But as alteration occurs on a microscale (5–40 μm), it is understandable that the mostly interstitial alteration products are not distributed homogeneously in the bulk sample.

The correlation of Rb/Sr ratios and contents of radiogenic ^{87}Sr (Fig. 5b) shows that the leached samples are characterized by uniform ^{87}Sr contents and that the ^{87}Sr concentrations of leach fluids decrease with leaching intensity ($D_3 - D_5 - D_7$), until the impact melt ^{87}Sr-level is reached. This can be taken as further evidence that the samples used in the age calculation can be regarded as being free of alteration components.

Nevertheless, Fig. 5c (according to a method described by Faure (1977)) unequivocally proves that a mixing relationship between impact melt (ideally sample B) and alteration component (D_3) exists, and, thus, that the regression line of Fig. 4 *also* represents a mixing-line. However, due to the above-given evident support for the validity of the 185 Ma age stating that the samples used for the age calculation are free of alteration products, and as some K-Ar ages (173 ± 8 and 165 ± 5 Ma) given by Lambert (1977) for cryptocrystalline impact melt samples from Rochechouart — evidently not treated for secondary alteration products — are reasonably close to our 185 Ma value, we believe that it has been sufficiently shown that this age should be taken as the best datum presently available for the crystallization age of the Rochechouart impact melt.

II. Mixing calculations and nature of meteoritic projectile

Despite the fact that considerable efforts were involved to achieve complete separation of feldspar and pyroxene, no important difference could be measured in Rb and Sr concentrations of the two mineral separates (Table II). The only possible explanation is that feldspar in 5/19 is K-feldspar (low Sr contents). Kraut (1969) has noted the existence of orthoclase crystals in Babaudu impact melt. Our X-ray diffraction analysis of coarse-grained feldspar sample 5/19 and of finer-grained feldspar from other Rochechouart impact melt samples showed that the total 42 vol % feldspar (Table I) consist of pure *high-sanidine* of Or_{100} composition. Feldspar (and tiny, < 5 μm, ilmenite crystals) are found to be the first crystallizing phases in the matrix of 5/19. SEM-EDAX studies showed that no traces of Na or Ca can be detected in 5/19 feldspar laths. Fig. 7b presents a $K_{K\alpha}$-distribution image of a typical area of a 5/19 thin section illustrating K-free pyroxene areas, K-rich sanidine, and mesostasis that contains about 50 % of the K content of sanidine (from comparison of impulse counts).

From Table III it is obvious that all impact melt samples from Rochechouart analysed to date are highly potassic in composition and lack significant amounts of Na or Ca. With the results of his previously mentioned two-component mixing model in mind, Lambert explains these alkali element abundances by the strong influence of weathering and hydrothermal alteration (Lambert, 1982). Having noted that the only feldspar phase of a rather fresh impact melt rock (5/19) from Rochechouart is K-feldspar, we performed mixing calculations in a five-component system trying to model the average impact melt

Table IV: Average Rochechouart target rock (mixture) and impact melt compositions (+ 1σ standard deviations) as used in HMX mixing calculations. In addition, K_2O, CaO, and Na_2O contents are given. Data (wt %) from Lambert (1977, 1982), Janssens et al. (1977), Palme et al. (1980), and from Reimold et al. (in prep.); n. d. = not determined.

Parameter	Components					Mixture
	gneiss	leptynite	granite	amphibolite	diorite	
SiO_2	66.39±2.85	75.01±3.15	73.02±2.99	53.60±2.48	59.52±3.00	66.50±2.53
TiO_2	.86± .16	.20± .09	.22± .22	.95± .61	.83± .60	.74± .09
Al_2O_3	16.30±1.65	13.67±1.50	14.91± .98	17.41±1.88	18.62±2.00	16.57± .93
Fe_2O_3	5.74±2.41	2.09±1.23	1.71±1.71	9.75±1.42	6.88±2.00	3.71±2.23
MgO	2.78± .45	.61± .30	.74± .81	6.44±2.48	4.09±2.50	1.09±1.09
K_2O	3.35±1.59	3.48±1.36	5.67±1.38	.76± .37	5.24±n. d.	
CaO	1.87± .87	1.26± .95	.79± .51	8.40± .85	1.79±n. d.	
Na_2O	2.92±1.07	3.61± .65	3.20± .92	3.53± .02	3.02±n. d.	

composition with proportions of all outcropping country rocks (gneiss, granite, leptynite, diorite, and amphibolite − Table IV), using the HMX mixing program.

The calculations were restricted to five parameters − SiO_2, TiO_2, Al_2O_3, total Fe as Fe_2O_3, and MgO, as the alkali elements and Ca concentrations of the impact melt are disputable. The average impact melt composition presented in Table III is only marginally different from the mixture composition used in mixing calculations and is modified by new data from Oskierski (1983). One might expect that Fe, Mg, and Ti abundances could have been affected by alteration, too, but the fact that the composition of least altered sample 5/19 (Table III) is consistent with the average impact melt composition within error limits made us confident with respect to the validity of our parameter choice.

Three different calculations were made: a five component run (I), and two four-component runs (II and III) deleting either the diorite or the amphibolite component of very similar compositions. The results are given in Table V. Discrepancy factors calculated as a measure for the reproducibility of the mixture composition when based on calculated component proportions are close to 1, demonstrating the high precision of the mixing calculation. Thus a probable target composition for the area of the Rochechouart meteorite crater of 66 ± 10 % gneiss, < 8 % leptynite, 26 ± 5 % granite, < 5 % amphibolite and 12 ± 5 % diorite can be predicted. It is interesting to note the HMX gneiss and granite proportions are well within the range of Lambert's (1982) two component model results. However, this study proves that an additional basic component of at least 10 % or a maximum of 18 % contributed to the impact melt source rock mixture.

From the INA analysis of four amphibolite aliquots the average abundances of siderophile elements Cr (255 ± 3 ppm), Co (29.4 ± 3 ppm), Ni (65 ± 20 ppm), Ir (2 ppb), and Au (1 ppb) are known. For the calculation of the Rochechouart indigenous component we had to assume that the dioritic rocks, for which no siderophile element analyses were available, are very similar in composition to the amphibolites and do not contain higher siderophile element contents. A comparison of our amphibolite data with trace element analyses of diorites compiled by Wedepohl (1978) supplies additional evidence justifying this assumption. Siderophile element data for typical gneisses and granites from the Haut-Limousin are given by Janssens et al. (1977) and Lambert (1977). The highest previously measured Ni contents of Rochechouart target rocks are 15 ppm in orthogneisses, and Co values are generally lower than 10 ppm. Clearly the introduction of a 15 % mafic com-

Table V: Results of HMX (Stöckelmann and Reimold, 1984) mixing calculations: modelling the impact melt mixture (proportions are given in %, DF = discrepancy factor).

component	run I	run II	run II
gneiss	62.9±9	76.4±8	57.5±10
leptynite	0±5	0±4	0± 8
granite	26.8±5	23.6±7	29.1±13
amphibolite	0±5	0±3	−
diorite	10.3±5	−	13.5± 8
DF	1.70	1.30	1.30

ponent of assumed amphibolitic composition raises the indigenous level well above those old values.

On the basis of the mixture proportions of Table V, indigenous correction (I.C.) factors of 12.7 ppm Co, 20.8 ppm Ni, 2.01 ppb Ir, and 1.8 ppb Au were determined. Cr abundances are only known for the amphibolite proportion, so that the value of 38 ppm was tentatively used as lower limit for the indigenous Cr contribution. With these values it was possible to make indigenous corrections of all impact melt analyses by Palme et al. (1980) and Janssens et al. (1977). These authors showed that inter-element ratios for Rochechouart impact melt samples are subject to considerable data scatter, probably due to non-negligible mobilization of certain elements such as Ni or Au by alteration processes (Lambert, 1982). Therefore we will only give average I.C.-corrected inter-element ratios (and 2 σ standard deviations) to describe the meteoritic siderophile component: Ni/Co = 28.0 ± 14, Ni/Ir = 36160 ± 28700, Ni/Cr = 1.29 ± 0.76. These corrected values definitely support the conclusion of Palme et al. (1980) that the Rochechouart projectile was of *chondritic* nature (compare e. g. chondritic ratios given by Palme et al., 1980 or Reimold, 1982).

III. Crystallization and Alteration Processes

The second purpose of the mixing calculations was to determine the processes responsible for the extreme alkali element concentrations (Table III) of the Rochechouart impact melt. On the basis of our proposed target mixture, the indigenous K_2O contribution should not be higher than 4.1 ± 1.6 wt % instead of the 10 wt % observed. Lambert (1982) discussed this discrepancy in detail; however, until now the common opinion was that excess K was added during chemical weathering. We do not deny the probable influence of chemical weathering on the bulk alkali element content, and we are conscious of the fact that addition of alkali elements due to weathering is a process frequently observed in nature (e. g. Reimold et al., in prep.). However, simple calculations show that a modal proportion of 42 vol % of sanidine (Or_{100}) (Table I) and of 20 vol % of potassic mesostasis require at least 8 wt % K_2O in the original melt to allow for completion of the observed crystallization process. This calculation is justified, as feldspar crystallization was restricted to the K-feldspar stability field − with Ca nevertheless present all the time, as the existence of late-crystallized Ca-pyroxene proves. Thus it is concluded that the maximum contribution of K_2O due to weathering can only be 3.7 wt %, whereas the rest of the excess K analysed in the impact melt and needed as soon as the crystallization process is set off − ∼4 wt % K_2O − can only be explained by assuming that selective melting of target rock feldspar highly enriched the impact melt in K in comparison to the target rock mixture. Preferred melting of feldspar is a process known from other impact melt sites (e. g. Reimold, 1982). In addition, support for the validity of this theory is the petrographical observation that more than 95 % of all mineral clasts observed in Rochechouart impact melt consist of quartz (with only 8 % leptynite component in the source rock mixture).

Our preferred explanation for the relatively large deficiency of the other alkali element Na in the impact melt (only 0.34 ± 0.3 wt % (Table III) against 3.4 wt % in the target rock mixture) is removal of this highly volatile element with the impact-generated gas phase during the early excavation of the crater (compare East Clearwater impact melt; Palme et al., 1979). Of course, weathering will affect the Na content even more than the K content (Krauskopf, 1979), but the removal of Na must have well proceeded before the impact melt crystallization was initiated, as no primary crystals of (K, Na)-feldspar can be detected.

Fig. 7a
SEM image of 5/19 impact melt (scale bar = 100 μm). The texture is determined by a subophitic network of large feldspar laths intergrown with pyroxene (e.g. at right margin). Interstices are filled with altered mesostasis (e.g. at left corners). Quartz fragments are very rare (center and at middle left rim).

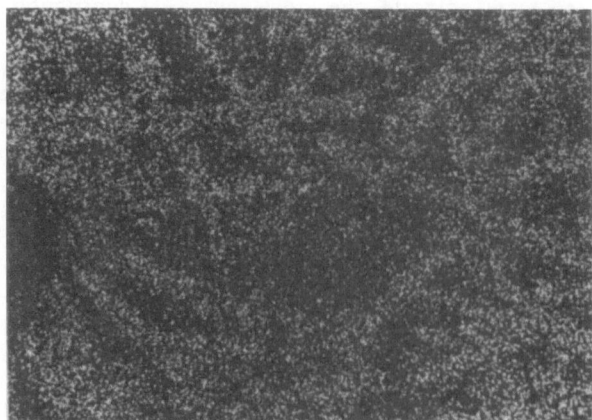

Fig. 7b
$K_{K\alpha}$-Xray distribution within same area as Fig. 7a. Clearly the sanidine crystals and K-rich mesostasis can be distinguished from pyroxene (right margin) and quartz clasts.

IV. Dike Breccias

Extensive petrographic descriptions of the four types of dike breccias (Fig. 8) cutting through the crater floor of the Rochechouart structure have been published by Oskierski and Bischoff (1983). The prominent characteristics and chemical compositions of selected samples are summarized in Table VI. As part of this work, mixing calculations were performed using the HMX method and basement rock compositions of Table IV in order to obtain the proportions of local and foreign dike-filling material. The results for four-component (gneiss, granite, leptynite, amphibolite components) calculations are given in Table VII.

Melt breccia of type 1-A is definitely different in composition from the coherent impact melt (compare Tables III and IV). As about 60 % of the dike filling is contributed from a

Fig. 8
Textural types of dike breccias from the Rochechouart subcrater basement. Widths of field of view are 3500 μm.

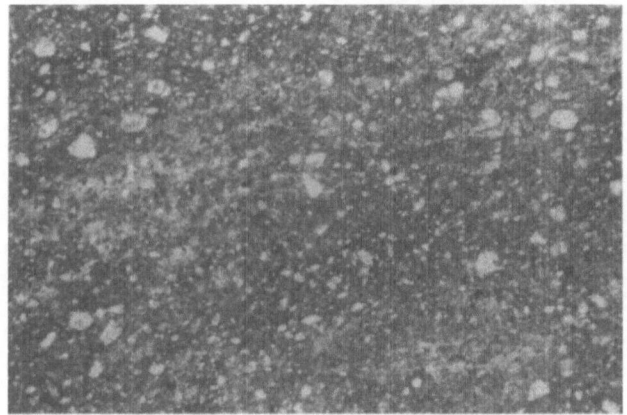

Fig. 8a
Impact melt breccia 1-A: note extremely fine grain size of matrix and absence of lithic clasts; a subdiagonal schlieren and clast orientation can be described.

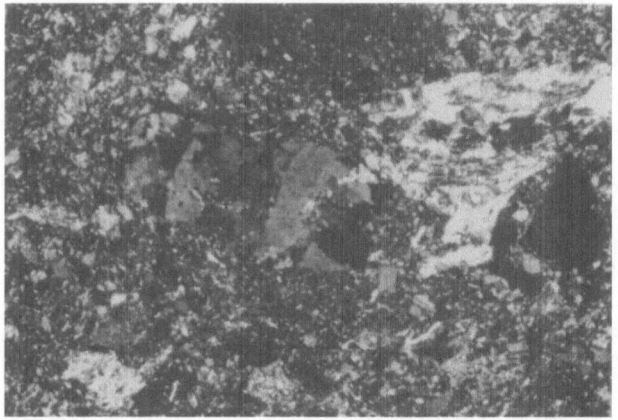

Fig. 8b
Impact melt breccia 1-B of comparatively coarse grain size and containing lithic clasts (center) of highly varied grain size.

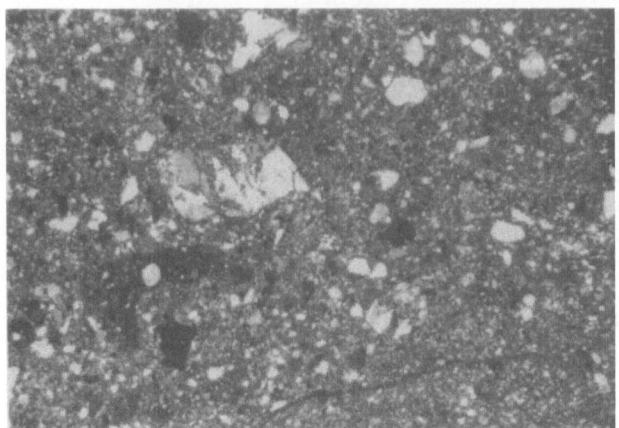

Fig. 8c

Clastic polymict breccia 2-A: larger clasts set into a rather fine-grained clastic matrix give the impression of subparallel orientation. The relatively few clasts are exclusively mineral fragments.

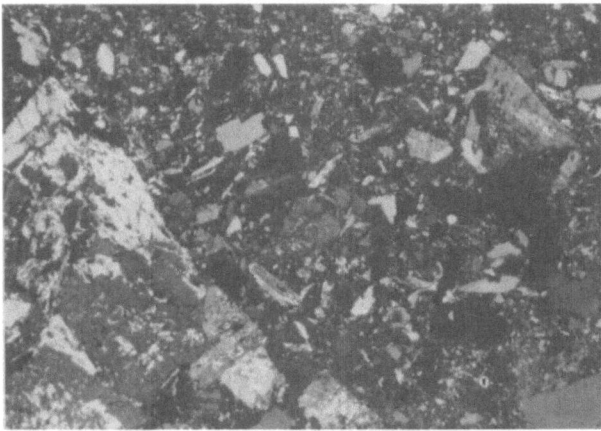

Fig. 8d

Clastic polymict breccia 2-B with distinct grain size distribution of clasts and a significantly higher clast content than observed in type 2-A breccias (cf. Fig. 8c). For more petrographic details, refer to the paper by Oskierski and Bischoff (1983).

mafic component (amphibolite, Table VII), while the only host rock of this dike type is gneiss, the logical conclusion is that these dike fillings were not formed of *local* material only. With respect to these observations, only one possible interpretation of the formation of type 1-A fillings can be suggested, namely that locally — *but not in situ!* — melted material was dynamically injected into these dikes prior to the completion of homogenization of the main impact melt layer. This hypothesis is also in good agreement with the petrographic features of type 1-A breccias; namely small clast sizes, flow structures in the matrix, and parallel orientation of clasts.

The chemical composition of *melt breccia 1-B* is very similar to the average impact melt composition. Also, the fillings of this variety are made up of several target rock proportions with gneiss and granite being major constituents (Table VII). The adequate conclusion would be that type 1-B fillings originate from the homogenized melt body and that these dikes were formed by gravitational emplacement of impact melt into opening fissures of the basement during the modification stage of the cratering process (during relaxation of the basement).

Table VI: Characteristics of the various dike breccia types found in the Rochechouart crater basement according to Oskierski (1983) and Oskierski and Bischoff (1983). Matrix in clastic breccias is defined as all material smaller than 10 μm in size.

type 1 (impact melt)		type 2 (clastic polymict breccias)	
variety 1-A	variety 1-B	variety 2-A	variety 2-B
width ≤ 1 cm	width up to dm	width from some mm up to some dm	maximum thickness: several m
complex geometry	simple dike forms	in general complex geometry	simple dike forms
no rock fragments	rock and mineral fragments		
clast sizes < 0.8 mm	clast size up to several cm	clast size up to several cm	clast sizes up to several cm
parallel texture: schlieren, subparallel clast orientation	no indication for dynamic injection	low fragment/ matrix ratio	high fragment/ matrix ratio
		flow structures	no indication for dynamic emplacement

Table VII: Results of mixing calculations for Rochechouart dike breccia fillings (data in %). For comparison the probable target rock composition calculated for coherent impact melt from the central crater region is presented as well.

dike type	gneiss	granite	amphibolite	host rock	sample No.
1-A	40.1	–	59.9	gneiss	5–11
1-B	41.3	58.6	–	gneiss	219592
2-A	–	100		granite	4–10
2-A	91.2	–	8.1	gneiss	5–11
2-A	97.8	0.03	2.0	gneiss	219594
2-B	42.4	57.5	–	granite	21059
2-B	53.8	46.1	–	granite	227413
2-B	53.1	46.8	–	gneiss	22941
2-B	84.5	14.0	–	gneiss	3–3

coherent impact
melt: 66 ± 10 % gneiss, < 8 % leptynite, 26 ± 5 % granite, < 5 % amphibolite, 12 ± 5 % diorite

Mixing calculations for various type *2-A clastic polymict breccia* fillings yielded consistent results: more than 90 % of the fillings consist of locally derived material with minor amounts of foreign components (samples 5–11 and 219594, Table VII). Thus two possibilities exist to interpret the 2-A formational processes: (i) type 2-A breccias are formed by in-situ fragmentation of basement rock, probably along shear planes, during the relaxation stage, while excavation of the crater is still in progress; (ii) the second possibility is that these dikes were formed by extensive mixing of abrased country rock and dynamically injected debris (short transportation ways) during the compressional early cratering phase.

Large amounts of foreign material have been added to the fillings of *clastic polymict breccias of type 2-B*. The maximum width of these dikes can be several meters, and their fillings are characteristically coarser-grained than those of type 2-A (Table VII). Type 2-B fillings are chemically homogeneous. Therefore it is understood that their fillings are made up of local material plus debris that was added from the crater interior during the post-compressional (opening of wide fissures) excavation phase, when the outward movements of material could mix and homogenize the clastic debris.

Summary and Discussion

1. The age of the Rochechouart impact event was precisely determined to be 186 ± 5 (1σ) Ma. Previously ages of only 14 impact events that produced crater structures larger than 20 km in diameter have been determined (e.g. Grieve, 1983). Grieve (1983) showed that the production rate of 20 km terrestrial impact structures was probably not uniform through the time span from 250 Ma ago to present time; the cumulative number of Phanerozoic craters larger than 20 km in diameter yields two populations when plotted against formation times — one for craters produced before 180 Ma ago, and another for those younger than 120 Ma. At present it can not be un-ambiguously assessed whether this result reflects a true change in cratering flux or is only due to different crater preservation states (Grieve and Robertson, 1979). It is obviously important to determine additional impact ages, especially of ages between 100 and 200 Ma ago, to arrive at a better understanding of the terrestrial crater production rate.

   Also, the ongoing discussion whether the meteorite flux onto the Earth was responsible for major biological extinction periods (e.g. Snowbird Conference, Geol. Soc. Am. Special Paper *190*, 1982) is largely dependent on reliable age determinations of impact events. Alvarez and Muller (1984) recently presented a correlation of mass-extinction times and crater production statistics and suggested that the extinction cycle of nearly 26 Ma and the impact cycle of about 24 Ma are the same within analytical uncertainties. Most interesting are their probability estimates that such agreement is improbable at the low rate of 1 to 10^3. Alvarez and Muller based this calculation on a K-Ar age previously suggested by Lambert (1977). Obviously such statistics *do* need new accurate data!

2. The formation of the impact melt of Rochechouart from various actually outcropping basement rock lithologies was established, a result necessary for better determinations of the nature of the *chondritic* projectile and geochemical considerations with respect to the roles of formational and weathering processes in determining the chemical composition of the impact melt.

3. To our knowledge this work presents the results of a first complete chemical and petrographical analysis of all breccia dike types occurring in the basement of a meteorite crater. On the basis of accurate mixing calculations, the derivation of the dike fillings could be extricated and placed in the cratering process. It remains to be tested now whether the proposed formational processes can be verified as well for breccia dikes from other impact craters.

Acknowledgements

The first author (W.U. Reimold) is grateful to Dr. B. Grauert for permission to perform the isotope analyses of this study at the Central Laboratory of Geochronology, Münster. Early versions of the manuscript were reviewed by H.L. Allsopp and B. Grauert, and we gratefully acknowledge the technical assistance of Mrs. F. Möllers and Mrs. F. Variava.

References

Alvarez, W. and R.A. Muller (1984): Evidence from crater ages for periodic impacts on the Earth. Nature 308, 718–720.

Bischoff, L. and W. Oskierski (1986): Fractures, pseudotachylite veins, and breccia dikes in the crater floor of the Rochechouart impact structure, SW France, as indicators of crater forming processes. (this volume)

Bogard, D.D., F. Hörz, P. Johnson and D. Stöffler (1981): Resetting of $^{40}Ar/^{39}Ar$ ages in suevite ejecta from the Ries Crater. Lunar Planet Sci. XII, 92–93.

Dence, M.R. (1965): The extraterrestrial origin of Canadian craters. Am. J. Sci. 123, 941–969.

Faure, G. (1977): Principles of isotope geology. J. Wiley and Sons, New York.

Grieve, R.A.F. (1983): Large-scale impact cratering on the terrestrial planets. Adv. Space Res. 2, 271–280.

Grieve, R.A.F. and P.B. Robertson (1979): The terrestrial cratering record. I. Current status of observations. Icarus 38, 212–229.

Horn, W. and A. El Goresy (1980): The Rochechouart Crater in France: stony and not and iron meteorite? Lunar Planet Sci. IX, 468–470.

Janssens, M.J., J. Hertogen, H. Takahashi, and E. Anders (1977): Rochechouart meteorite crater: identification of projectile. J. Geophys. Res. 82, 750–758.

Jarrar, G., A. Baumann and H. Wachendorf (1983): Age determinations in the Precambrian basement of the Wadi Araba area, southwest Jordan. Earth Planet. Sci. Lett. 63, 292–304.

Jessberger, E.K. and W.U. Reimold (1980): A late cretaceous age of the Lappajärvi impact crater, Finland. J. Geophys. 48, 57–59.

Jessberger, E.K., T. Staudacher, B. Dominik, T. Kirsten and O.A. Schaeffer (1978): Limited response of the K-Ar system to the Nördlinger Ries giant meteorite impact. Nature 271, 338–368.

Krauskopf, K.B. (1979): Introduction to geochemistry. McGraw-Hill Book Company, New York.

Kraut, F. (1969): Über ein neues Impaktitvorkommen im Gebiet von Rochechouart-Chassenon (Départements Haute-Vienne und Charente, Frankreich). Geologica Bavarica 61, 428–450.

Kroll, H. und D. Stöckelmann (1979): Harmonische Least-Squares Datenanalyse. Fortschr. Miner. 57, Beiheft 1, 74–75.

Lambert, P. (1974): Etude géologique de la structure impactitite de Rochechouart (Limousin, France) et son contexte. Bull. BRGM Sec. I, 3, 153–164.

Lambert, P. (1977): Les effets des ondes de choc naturelles et artificielles et le cratère d'impact de Rochechouart (Limousin France). Ph.D. Thesis, Université Paris-Sud.

Lambert, P. (1982): Anomalies within the system Rochechouart target rock-meteorite. Geol. Soc. Am., Special Paper 190 (Snowbird Conference).

McKinley, J.P., G.J. Taylor, K. Keil, M.-S. Ma, and R.A. Schmitt (1981): The origin of Apollo 16 dimict breccias. Lunar Planet. Sci. XII, 691–692.

McKinley, J.P., G.J. Taylor, K. Keil, M.-S. Ma, and R.A. Schmitt (1982): Apollo 16 impact melt sheets. Lunar Planet. Sci. XIII, 496–497.

McKinley, J.P., G.J. Taylor, and K. Keil (1983): White portions of Apollo 16 dimict breccias are polymict. Lunar Planet. Sci. XIV, 483–484.

Oskierski, W. (1983): Geologisch-Petrographische Untersuchungen im Zentralbereich der Impaktstruktur von Rochechouart, SW-Frankreich, unter besonderer Berücksichtigung der Petrographie und Geochemie von Brecciengängen des Krater-Untergrundes. Diploma Thesis, Univers. Münster.

Oskierski, W. and L. Bischoff (1983): Petrographic, geochemical, and structural studies on impact breccia dikes of the Rochechouart impact structure, SW France. Lunar Planet. Sci. XIV, 584–585.

Palme, H. (1980): The meteoritic contamination of terrestrial and lunar impact melts and the problem of indigenous siderophiles in the lunar highland. Proc. Lunar Planet. Sci. Conf. 11th, 481–506.

Palme, H., E. Göbel, and R.A.F. Grieve (1979): The distribution of volatile and siderophile elements in the impact melt of East-Clearwater (Quebec). Proc. Lunar Planet. Sci. Conf. 10th, 2465–2492.

Palme, H., W. Rammensee and W.U. Reimold (1980): The meteorite component of impact melts from European impact craters. Lunar Planet. Sci. XI, 848—850.

Rehfeldt, A. (1983): Petrographische und chemische Untersuchungen an Gangbreccien im kristallinen Krateruntergrund des Nördlinger Ries. Diploma Thesis, Univers. Münster.

Reimold, W.U. (1982): The Lappajärvi meteorite crater, Finland: petrography, Rb-Sr, major and trace element geochemistry of the impact melt and basement rocks. Geochim. Cosmochim. Acta 46, 1203—1225.

Reimold, W.U., R.A.F. Grieve and H. Palme (1981): The Rb-Sr age of the impact melt from East Clearwater, Quebec. Contrib. Mineral. Petrol 76, 73—76.

Reimold, W.U., J. Nieber-Reimold, W. Oskierski, and A. Rehfeldt (1983): A geochemical and chronological study on amphibolites and granitic rocks from the Haut-Limousin, Massif Central. Fortschr. Miner. 61, Beiheft 1, 178—180.

Reimold, W.U., W. Oskierski, and J. Huth (in prep.): Petrography, REE and Rb-Sr isotopic systematics of amphibolitic and granitic rocks from the Haut-Limousin, France.

Stöckelmann, D. and W.U. Reimold (1984): The HMX mixing calculation program, version 1984. Institute of Mineralogy, Münster, FRG.

Stöffler, D. (1977): Research drilling Nördlingen 1973: Polymict breccias, crater basement, and cratering model of the Ries impact structure. Geologica Bavarica 75, 443—458.

Wedepohl, K.H. (ed.) (1969): Handbook of Geochemistry. Springer-Verlag Berlin—Heidelberg— New York.

Wilshire, H.G., T.W. Offield, K.A. Howard and D. Cummings (1972): Geology of the Sierra Madera Cryptoexplosion structure, Pecos County, Texas, U.S. Geol. Surv. Prof. Paper 559-H.

The Use of Airborne and Spaceborne Radar Images for the Detection and Investigation of Impact Structures

Barbara Theilen-Willige
Bergische Str. 7, D-4320 Hattingen 16, FRG

Key Words

Detection and investigation of impact craters
SLAR-X-band Projeto RADAMBRASIL images
Shuttle Imaging Radar (SIR-A) L-band images
Comparative analysis of LANDSAT images

Abstract

This paper illustrates the application feasibilities of airborne and spaceborne radar images for the detection and delineation of impact structures.

Especially in tropical and subtropical environments that are often obscured on aerial photographs and LANDSAT images by cloud cover or by a dense vegetation, SLAR (Side Looking Airborne Radar) images have become a useful tool for reconnaissance work and have contributed to the detection and to the geomorphologic and geologic investigation of impact structures. The example from the complex Araguainha impact structures in Central Brazil is summarized here to demonstrate the value of SLAR Projeto RADAMBRASIL X-band radar mosaics (1976) for the investigation of this structure.

In arid and semi-arid areas, spaceborne SAR (Synthetic Aperture Radar) L-band images acquired by the Space Shuttle Columbia in November 1981 have clearly demonstrated their value for the detection of still unknown geologic features due to the susceptibility of radar signals to the micro- and macro-relief and to the capability of radar signals to penetrate into dry sedimentary covers up to several meters. A large number of circular structures covered by thin sedimentary layers have become visible on the SIR-A (Shuttle Imaging Radar) images because of this penetration capability, among them circular features with impact origin.

Examples of known and suggested impact craters are demonstrated from North Algeria, Southeast Syria and North Saudi Arabia.

1. Introduction

In the past 20 years a large number of impact structures has been discovered by interpretations of aerial photographs and of LANDSAT images.

The objective of this report is a contribution to the knowledge of terrestrial impact structures by the use of airborne and spaceborne SAR (Synthetic Aperture Radar) systems: the SLAR (Side Looking Airborne Radar) images of the Projeto RADAMBRASIL-project (1972–1976) and the spaceborne SIR-A (Shuttle Imaging Radar) images acquired by the Space Shuttle Columbia in November 1981.

2. Interpretation of a SLAR-X-band Projeto RADAMBRASIL radar Mosaic from the Complex Araguainha Structure in Central Brazil

The geomorphologic and geologic evaluation of airborne radar images is a useful tool for the detection and investigation of impact craters situated in remote areas, often obscured by clouds and a dense vegetation cover.

One example from the complex Araguainha impact structure is summarized here to demonstrate the application feasibilities of SLAR-Projeto RADAMBRASIL-radar mosaics in this poorly accessible subtropical region that is well suited for radar investigations.

Technical data of the SLAR Projeto RADAMBRASIL images are listed in Table I.

The Araguainha structure is a complex, multiringed impact structure 40–50 km in diameter, located at the border of Southwest Goiás and Southeast Mato Grosso/Central Brazil (latitude: 16°30′ to 17°15′ south; longitude: 52°45′ to 53°15′ west), see Fig. 1. It has been described recently by Theilen-Willige (1981, 1982) and Crosta et al. (1981, 1982) indicating its origin by a cosmic impact.

The Araguainha structure is geomorphologically characterized by a large ring-shaped basin, depicting a central elevation, ring depressions, annular arranged ridges, arcs of isolated hills, terraced walls and distinct escarpments at the border.

The central uplift consists of a crystalline nucleus 3 km in diameter that contains hypidiomorphic and partly porphyritic rocks with alkali-granitic to alkali-syenitic character. The crystalline nucleus is surrounded by an annulus of mixed, polymictic breccias, suevites and tilted Paleozoic sediments, predominantly sandstones, conglomerates and siltstones (see Fig. 2).

Table I: SLAR-Projeto RADAMBRASIL characteristics according to Lima (1976) and Greeves et al. (1975)

wavelength (cm)	: 3.12 (X-band)
look direction	: W
depression angle	: 45° − 13°
polarization	: HH
swath width (km)	: 37
resolution (m)	: 16
altitude (m)	: 11 000
flight direction	: N-S, spaced 15 minutes of longitude
scale of the semi-controlled radar-mosaic Goodyear Electronic Mapping System 102	: 1:250 000

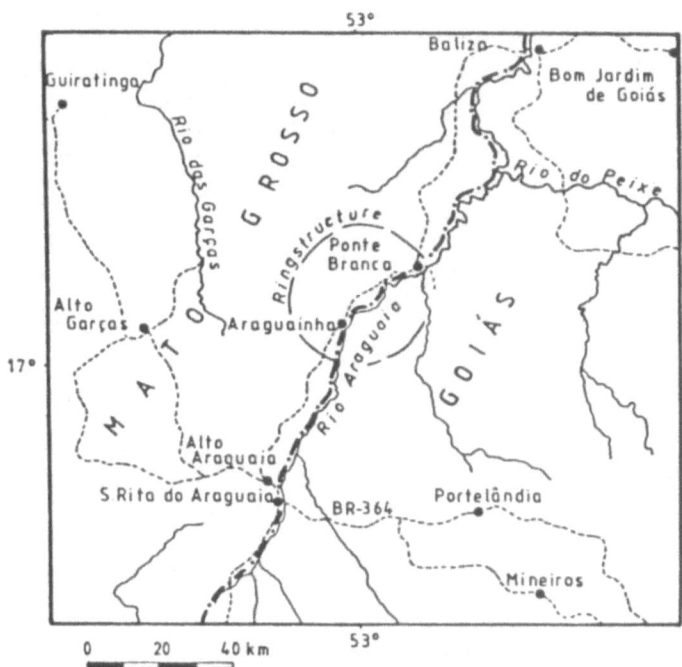

Fig. 1:
Index map of the Araguainha structure

The LANDSAT image of this area and its enhancement by digital techniques of image processing is of great value for the detection and mapping of lithologic units (see Plates 1a and b).

The radar mosaic of the Araguainha structure (see Plate 2), however, demonstrates clear advantages for the geomorphologic and structural evaluation.

2.1. Geomorphologic Evaluation of the Radar Mosaic

Due to the distinct sensitivity of radar signals to micro- and macro-relief, the SLAR images enhance geomorphologic details. The radar shadows emphasize topography and provide a clear relief impression of three-dimensionality. The detailed terrain information helps to identify the landform features of the Araguainha structure. Although weathering and erosion processes have strongly modified the original impact structure (the probable Mesozoic age of the Araguainha structure is still in discussion, Crosta, 1983, pers. comm.), the multiringed basin character is clearly visible on the radar mosaic.

Based on the radar mosaic, a geomorphologic sketch map has been elaborated (see Fig. 3). The existence of the identified geomorphologic units on the radar mosaic is confirmed by the hypsometric map (see Fig. 4) as well as by stereoscopic interpretations of aerial photographs and by field investigations (Theilen-Willige, 1981). The pentagon-like central uplift with the central basin and the inner and outer ring of hills is clearly detectable (compare Plate 2 and Fig. 3). the topographic map indicates heights of more than 600 m (highest point: 699 m) in the northern part of the outer ring of tilted Furnas sandstones.

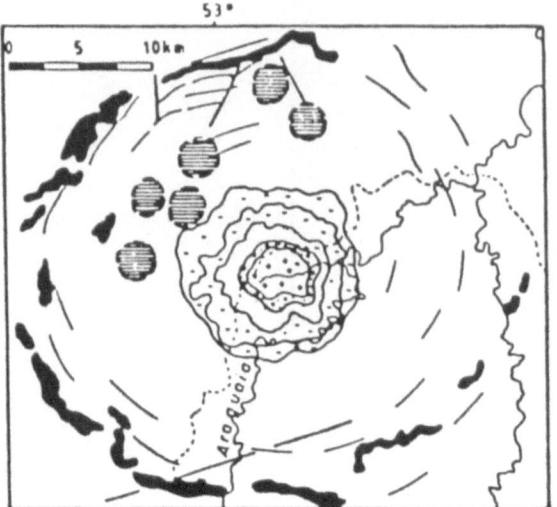

Fig. 2
Geologic sketch-map of the Araguainha
structure based on field observations,
interpretation of the LANDSAT
scene — 1089—13005 (Theilen-Willige,
1981) and Geologic Maps (Silveira &
Ribeiro, 1971; Schobbenhaus et al.,
1975)

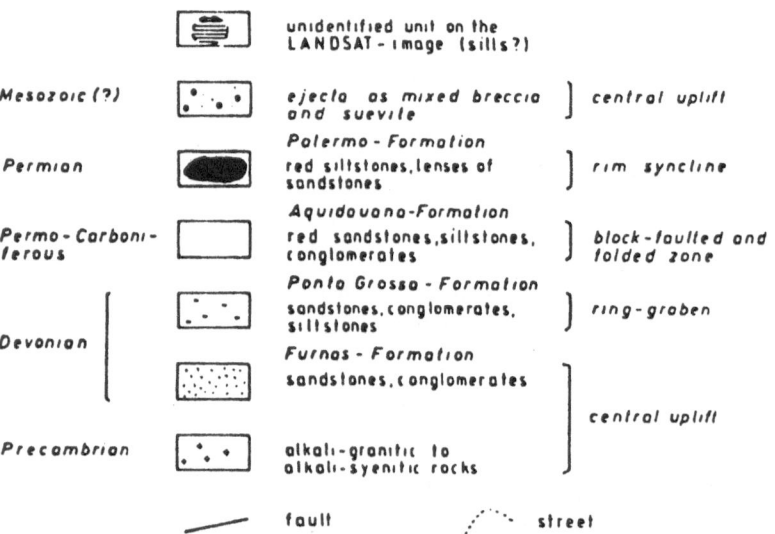

The ring depression surrounding the central uplift depicts height levels between 470 to 490 m. The various ring walls generally reach altitudes between 500 to 600 m. The heights of the plateaus constituted of Paleozoic sedimentary rocks surrounding the impact basin range from 800 to 900 m. The radar mosaic indicates a basin diameter of 50 km.

Because of the synoptic view of the Araguainha structure on the radar mosaic, some geomorphologic features (as for example the ring walls) are better detectable on the radar mosaic than on aerial photographs.

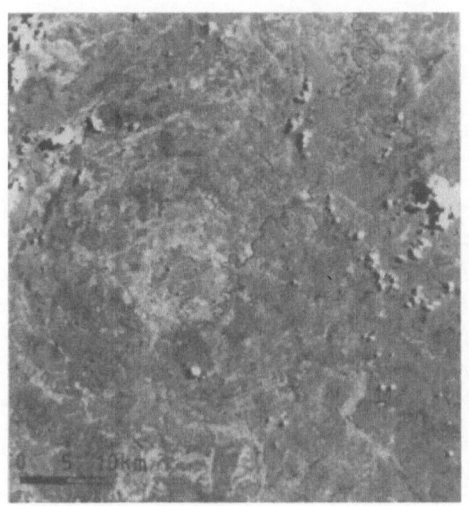

Plate 1a
LANDSAT scene (MSS-7, No. 1089-13005)
from the Araguainha structure

Plate 1b
LANDSAT-MSS-5 scene from the
Araguainha structure

⟸ illumination

Plate 2:
SLAR X-band Projeto RADAMBRASIL semicontrolled radar mosaic from the Araguainha structure
Folha SE.22-V-G, Projeto RADAMBRASIL, 1976

2.2. Structural Investigation of the Radar Mosaic

The radar mosaic of the Araguainha structure contains an amount of structural information related to the impact event, due to the shadowing characteristics of SLAR images
that enhance topographically expressed linear and curvilinear features such as faults and
curvillinear traces of the impact event (see Fig. 5). The low resolution (16 m) of the radar

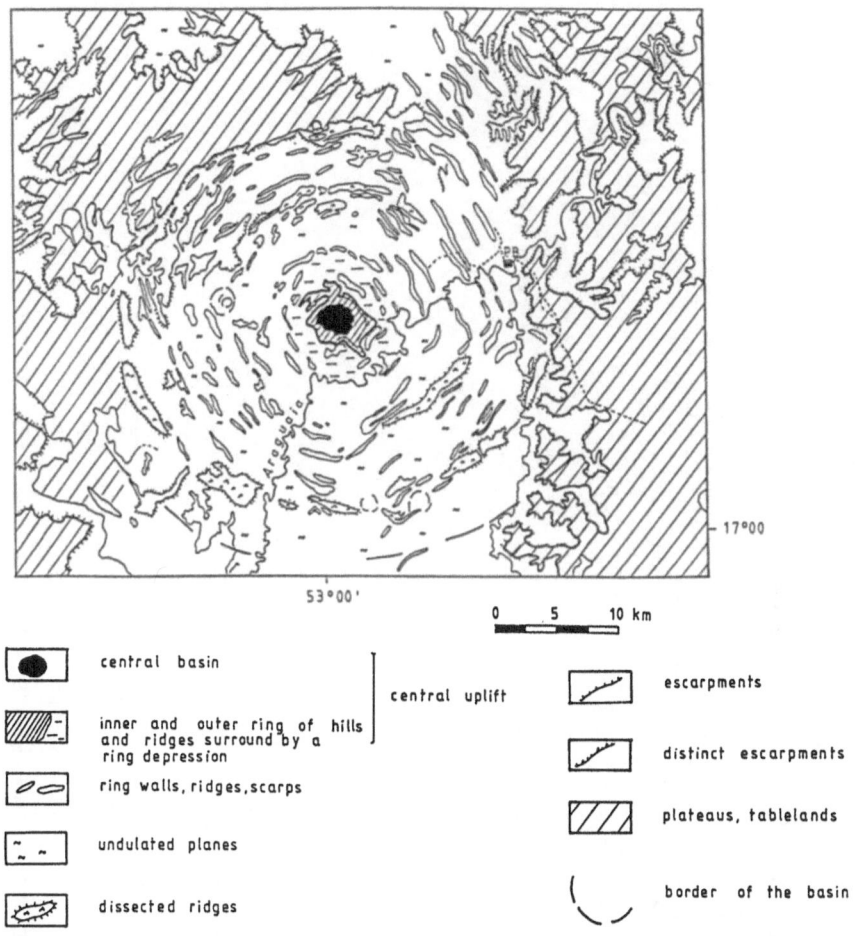

central basin

inner and outer ring of hills
and ridges surround by a
ring depression

central uplift

ring walls, ridges, scarps

undulated planes

dissected ridges

escarpments

distinct escarpments

plateaus, tablelands

border of the basin

Fig. 3:
Geomorphologic sketch-map from the Araguainha structure based on the SLAR Projeto RADAM-BRASIL radar mosaic, Folha SE.22-V-C, Projeto RADAMBRASIL, 1976 (P.B. = Ponte Branca)

images tends to suppress vegetative detail and to emphasize topographic details reflecting the underlying geology. The structural evaluation of the radar mosaic is primarily based on the topographic expression of structural features and secondarily on the drainage pattern and tonal anomalies.

The look direction (to W) of the radar is of great importance for the detection of linear features. (MacDonald, 1969). (The look direction is the direction perpendicular to the N-S flight direction of the aircraft.) Linear and curvilinear features oriented perpendicular (N-S) to the look direction are enhanced and those oriented parallel (E-W) to the look direction are suppressed on the radar images (see Fig. 5). The radar mosaic traces radial and tangential faults of the Araguainha structure. The curvilinear features obviously reflect the courses of shockwaves in the rocks. Most of the curvilinear features are distributed in a concentric pattern. Sometimes they interfere which might be explained by inhomogeneities in the subsurface during the impact event.

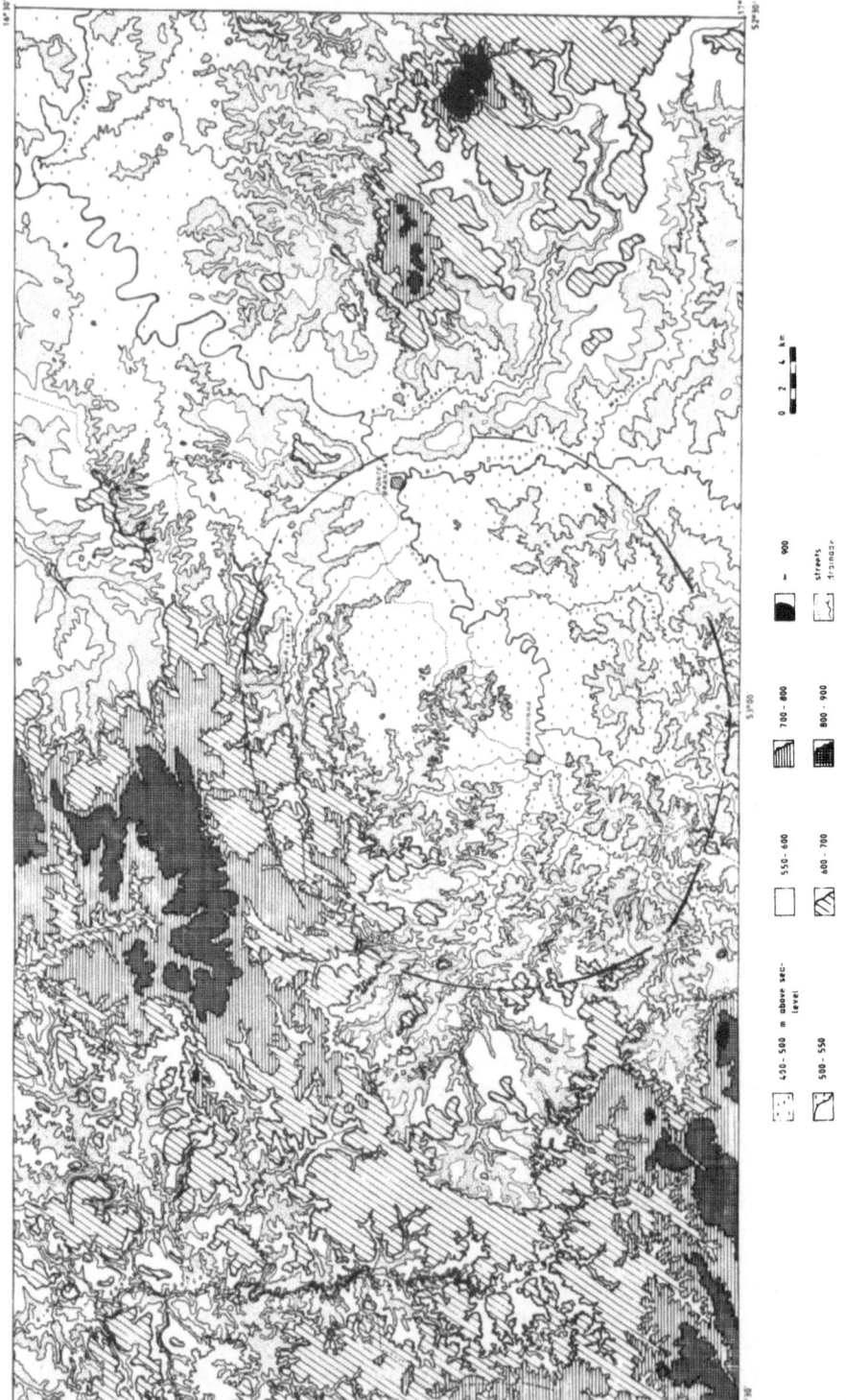

Fig. 4

Hypsometric map of the Araguainha structure
Based on the Topographic Maps "Carta do Brasil", Folha SE.-22-V-A-V, Araguainha, SE.-22-V-A-VI, Ponte Branca, 1:100 000, IBGE, 1976

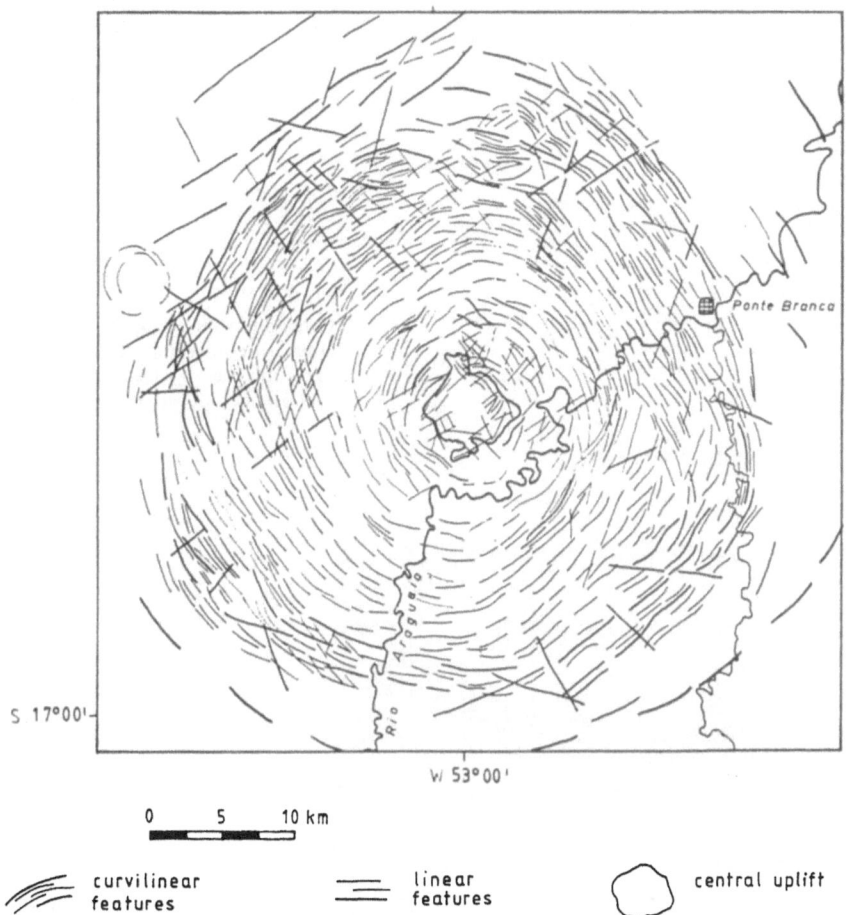

Fig. 5:
Structural evaluation of the SLAR mosaic from the Araguainha structure

3. Interpretation of the SIR-A L-band image and of the LANDSAT scene from the simple Talemzane crater in North Algeria

In November 1981 the Space Shuttle Columbia carried its first scientific payload into earth orbit. Part of this payload was the Shuttle Imaging Radar (SIR-A) which acquired radar images of about 10 million square kilometers around the world (Elachi et al., 1982). The SIR-A characteristics are described in Table II. As an example for the use of SIR-A images for the detection and investigation of impact craters, the Talemzane crater in North Algeria is represented (see Plate 3 and Fig. 6).

Variations in tones of the SIR-A image are caused by changes in radar backscatter that are a function of physical properties of the terrain.

Fig. 6
Location of the simple Talemzane crater and other impact and impact-like structures in Algeria according to Lambert et al. (1981)

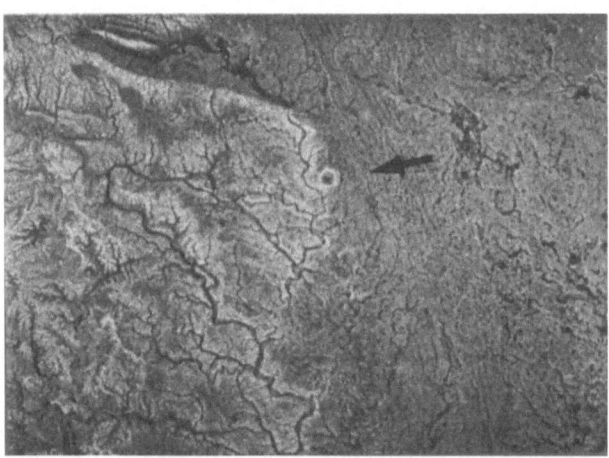

Plate 3
SIR-A image from the bowl-shaped Talemzane crater in North Algeria

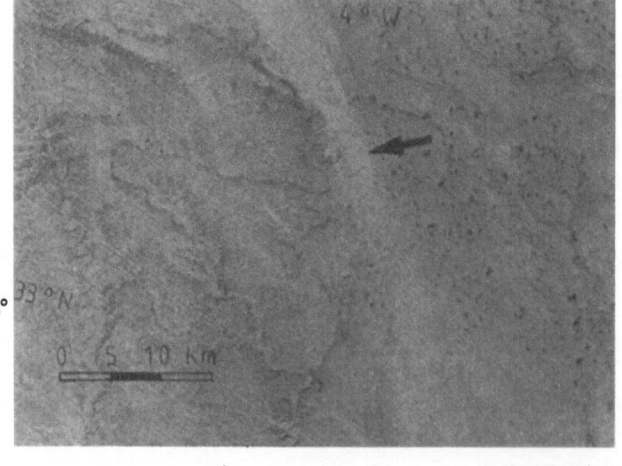

Plate 4
LANDSAT scene (MSS-7, No. 2396-09334) from the Talemzane crater

Table II: SIR-A characteristics according to Elachi (1982) and
Ford et al. (1983)

wavelength (cm)	: 23.5 (L-band)
look angle	: 47° ± 3°
Incidence angle	: 50° ± 3°
polarization	: HH
swath width (km)	: 50
resolution (m)	: 38
image data scale	: 1:500 000
average altitude (km)	: 260

The flat Hamada surfaces consisting of Tertiary sediments (right part of the image) appear in dark tones on the radar image (see Plate 3), because the incident radar waves had been reflected mainly away from the receiving antenna. The Cretaceous sediments (left part of the image) scatter the radar energy in all directions because of their strongly expressed micro- and macro-relief. Due to this diffuse scattering, they appear in light tones on the SIR-A image. At the upper center of the image the bowl-shaped Talemzane crater is clearly visible (see arrow). This crater was dated at about 0.5 to 2 million years old (Lambert et al., 1980). The Talemzane crater 1.7 km in diameter and 70 m deep is characterized by a rim of ejecta which is expressed on the radar image as a light ring. Field observations (Lambert et al., 1980) indicated that blocks up to a size of 1 m are not uncommon within these ejecta blankets. This blocky material appears as a diffuse reflector for the incident waves and is mainly responsible for the clear appearance of the Talemzane crater on the radar image. On the corresponding LANDSAT scene, the Talemzane crater nearly remains invisible (see Plate 4).

4. Interpretation of the SIR-A Image and of the LANDSAT Scene from the Complex Treibil Crater and Simple Tleachman Crater in Southeast Syria

As next examples for impact craters detected by the use of LANDSAT- and SIR-A images, two impact structures in Southeast Syria are represented. The complex Treibil crater 10 km in diameter (latitude: 32°15'N; longitude: 39°00'E) and the bowl-shaped Tleachman crater 4 km in diameter (latitude: 32°30'N; longitude: 39°25'E) have been mapped first by Kruck et al. (1981) based on LANDSAT interpretations. These obviously recent impact craters are located within lower Tertiary marls and limestones.

Plate 6 shows the SIR-A image of the two impact craters and their environment. On the radar image the Treibil crater is nearly invisible, because serveral houses have been built exactly at the center of the crater (see arrow 2, Plate 6). The building of these houses led to the planation and removal of ejecta material and impact traces. The Treibil crater is better detectable on the LANDSAT scene, see Plate 7. The Tleachman crater shows a light annulus on the radar image that can be related to the ejecta material (see arrow 1, Plate 6). The Tleachman crater appears very distinct on the radar image and less expressed on the LANDSAT scene.

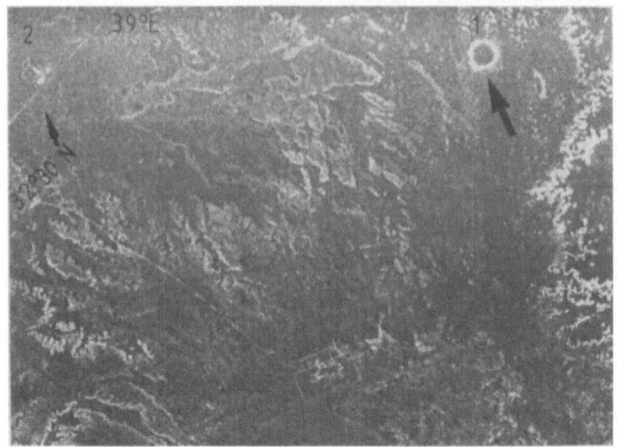

Plate 6
SIR-A image from the bowl-
shaped Tleachman crater (1)
and the complex Treibil crater
(2) in South-east Syria

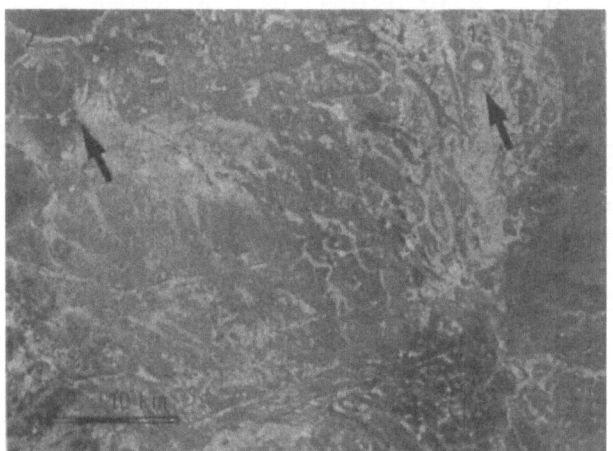

Plate 7
LANDSAT scene (MSS-5,
No. 2156-07191) from the
Tleachman crater (1) and
Treibil crater (2) in Southeast-
Syria

5. Interpretation of the SIR-A Image and the LANDSAT Scene from a Circular, Impact-like Structure near Al Hashima in North Saudi Arabia

Due to the penetration capability of radar signals to pierce dry sedimentary covers such as eolian and fluvial sand sheets up to several meters (McCauley et al., 1982), subsurface features become visible on radar images. The electromagnetic energy incident on the surface of dry sand sheets is partly reflected away and partly transmitted (and refracted) through the medium until being diffusely scattered from a rough substrate. This fact is of great importance for those areas where impact craters have been covered by thin sedi-

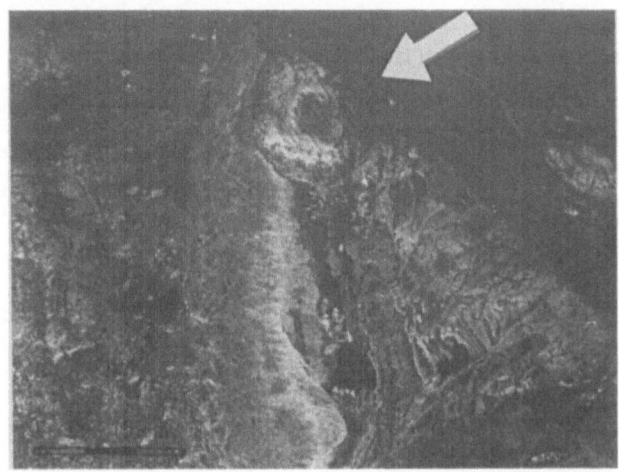

Plate 8
SIR-A image from a circular,
impact-like structure in North
Saudi Arabia

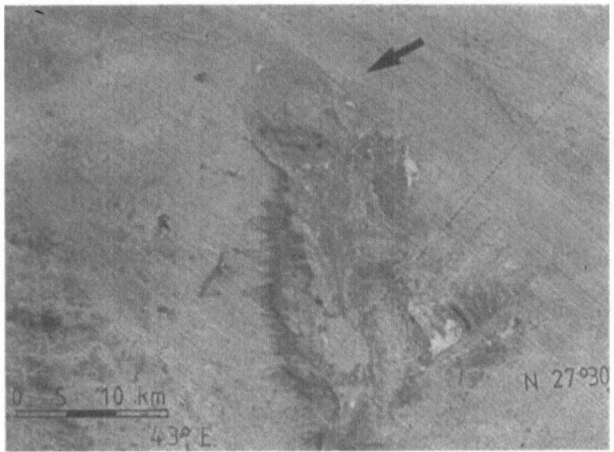

Plate 9
LANDSAT scene (MSS-5,
No. 1156-07094) from the
buried circular feature and its
environment

mentary layers as shown by the following example. Plate 8 shows a SIR-A image from a sircular, impact-like feature near the town Al Hashimah in North Saudi Arabia (latitude: $27°30'-28°00'$N; longitude: $43°00'-43°30'$E, image frame). On the SIR-A image appears a distinct circular structure 15 km in diameter (see arrow, Plate 8) which could be interpreted as a complex impact crater. At the center a central uplift is indicated by a light ring. Its northern part is covered by flucial and eolian sediments appearing black on the radar image.

On the corresponding LANDSAT scene (see Plate 9), Quaternary eolian and fluvial sediments are prevailing in light tones. Triassic and Devonian sediments are visible as dark-toned ridges. The arrow (Plate 9) demonstrates the area of the circular feature which is buried by sediment sheets.

6. Interpretation of the SIR-A Image and the LANDSAT Scene from a Circular, Impact-like Structure at the Northwest Coast of Saudi Arabia

Another example for a circular structure covered by eolian and fluvial sand sheets is located at the northwest coast of Saudi Arabia (latitude: 24°00′−24°30′N; longitude: 38°00′−38°30′E, image frame).

Plate 10 shows the SIR-A image of a circular structure situated near Yanbu'al Bahr-harbour. Quaternary marine, fluvial and eolian deposits appear in dark tones on the radar image whereas Precambrian rocks such as greenschists, diorites, granites, syenites and gneisses are recognized in light tones (USGS, 1963, Geologic Map of the Arabian Peninsula). The ring structure up to 15 km in diameter (see arrow, Plate 10) is described on the Geologic Map (1963) as a diorite-intrusive complex. Its circular outline and the concentric structural traces, however, resemble a deeply eroded complex impact crater. On

Plate 10
SIR-A image from a circular, impact-like feature on the northwest coast of Saudi Arabia

Plate 11
LANDSAT scene (MSS-5, No. 1159-07274) from a circular, impact-like structure (1) on the northwest coast of Saudi Arabia

the corresponding LANDSAT scene (Plate 11), the Quaternary sediments are visible in light tones. The arrow indicates the area of the circular feature visible on the SIR-A image. The circular feature is nearly completely covered by sedimentary deposits appearing in light tones on the LANDSAT image.

7. Conclusions

Airborne SLAR X-band images have shown their value for the geomorphologic and structural investigation of impact structures, especially in tropical and subtropical environments. Relatively detailed information about the relief and topographically expressed structural features can be derived by the evaluation of radar images.

Spaceborne L-band images are very useful for the detection of circular features, especially in arid to semi-arid areas due to the penetration capability of radar signals into dry sediments up to several meters. Even under sedimentary covers, buried ring structures have become visible on SIR-A images, among them impact-like structures.

It will be of great interest to explore SIR-B data that will be obtained in September 1984 by the Space Shuttle Columbia, concerning the detection of circular features which might represent impact craters.

Acknowledgements

The author would like to thank Dr. A. Sieber, Deutsche Forschungs- und Versuchsanstalt für Luft- und Raumfahrt (DFVLR), Oberpfaffenhofen and the colleagues from the Bundesanstalt für Geowissenschaften und Rohstoffe (BGR), Hannover for providing SIR-A- and LANDSAT-data. This support is gratefully acknowledged.

References

Crosta, A.P., J.C. Gaspar e M.A.F. Candia, 1981: Feicões de Metamorfismo de Impacto no Domo de Araguainha. Rev. Bras. de Geociências, 11, 3, 139–146. São Paulo

Crosta, A.P., 1982: Estruturas de Impacto no Brasil: Uma Sintese do conhecimento Atual. Anais Do XXXII Congr. Bras. de Geologia, Salvador/Bahia. V. 4, 1372–1377.

Elachi, C., W.E. Brown, J.B. Cimino, T. Dixon, et al., 1982: Shuttle Imaging Radar Experiment. Science, 218, 3 Dec., 996–1003.

Elachi, C., 1982: The Shuttle Imaging Radar (SIR-A) – Sensor and Experiment. IEEE Digest, Vol. II, IGARSS '82, Fa-6, 5.1–5.6, June 1–4. 1982, Munich.

Greeves, R.G., A. Anson, and D. Landon, 1975: Manual of Remote Sensing, Vol. 1, American Society of Photogramm.

Kruck, W., R. Rajab and W. Wagner, 1981: Geologic Map of the Hamad Basin-Project, Sheet 4, ACSAD, GTZ, BGR, Damascus, Hannover.

Lambert, P., J.F. McHone, Jr., R.S. Dietz and M. Houfani, 1980: Impact and Impact-like Structures in Algeria, Part I: Four Bowl-shaped Depressions. Meteoritics 15, 157–179.

Lambert, P., J.F. McHOne, Jr., R.S. Dietz, M. Briedj and M. Djender, 1981: Impact and Impact-like Structures in Algeria, Part II: Multi-ringed Structures. Meteoritics, 16, 3, 203–227.

Lima, M.I.C., 1976: Modelos Radargráficos de Estruturas Circulares na Regiao Amazónica. XXIX Congr. Bras. de Geologia, Ouro Preto/Minas Gerais.

MacDonald, H.C., 1969: Geologic Evaluation of Radar Imagery from Darien Province, Panama. Mod. Geology, 1, 1–63.

McCauley, J.F., S.G. Schaber, C.S. Breed, M.J. Grollier, et al., 1982: Subsurface Valleys and Geoarcheology of the Eastern Sahara Revealed by Shuttle Radar. SCIENCE, 218, 3 Dec., 1004–1020.

Theilen-Willige, B., 1981: Fernerkundungsverfahren bei geomorphologischen und strukturellen Untersuchungen an Intrusivkomplexen und an der Ringstruktur von Araguainha. Clausthaler Geowiss. Diss., 8, 210 pp., Clausthal-Zellerfeld.

Theilen-Willige, B., 1982: The Araguainha Astrobleme/Central Brazil. Geol. Rundschau, 71, 1, 318–327.

Schobbenhaus, C., K. Oguino, C.L. Ribeiro, L.A. Oliva, J.T. Takanohashi, 1975: Carta Geológica do Brasil ao Milionésimo, Folha Goiânia (SE-22), DNPM, Brasilia.

Silveira Filho, N.C. & C.L. Ribeiro, 1971: Informacões Geológicas Preliminares Sobre a Estrutura Vulcânica de Araguainha M.T., DNPM, Relatório Interno Inédito.

USGS, 1963: Geologic Map of the Arabian Peninsula, Map I-270 A, Scale 1:2 000 000.

The Hico Impact Structure of North-Central Texas

Leanne Wiberg Milton*
Department of Geology, Texas Christian University, Fort Worth, TX 76129, USA

Key Words

Impact structure
Circular disturbance
Shatter cone
Central uplift
Ring graben
Ring fault

Abstract

The Hico Structure is a pre-Pleistocene, post-Cretaceous impact structure centered 3 km north of the town of Hico in north-central Texas. Less eroded sections of the feature faithfully preserve an original 3 km-diameter "uplift surrounded by ring graben" pattern typical of impact structures larger than one kilometer. A shatter cone has been found within the central uplift. A larger circular anomaly 9 km in diameter surrounding the 3 km-diameter Hico Structure is visible on Landsat imagery, although no corresponding structural disturbance has been recognized in outcrop.

The central uplift of the Hico Structure consists of a plug of folded Cretaceous strata broken into irregular fault-bounded blocks from 25 to 200 m across. The intensity of the folding within the uplift increases toward the center. It is likely that the oldest units near the center have moved upward at least 80 m. A ring graben surrounds the central uplift. Differential downdrop of at least 18 m and 30 m has occurred along segments of the outer fault bounding the ring graben.

The subsurface extent of deformation just outside of the ring graben is shallower than 150–200 m below the present ground surface. Gravity and magnetic profiles of the structure reveal no excess or deficient subsurface mass.

The gross structural similarity to accepted impact structures and the presence of a shatter cone indicate an impact origin for the Hico Structure. Among accepted impact structures, the central uplift of the Hico Structure most closely resembles the central uplift of Sierra Madera. Both are strongly folded with gentle undulating folds on uplift flanks and tighter chevron-type folds inward.

---

* Present address: 1446 Crowell Road, Vienna, VA 22180, USA

1. Location and Geologic Setting

A pre-Pleistocene, post-Cretaceous circular structure is centered 3 km north of the town of Hico, Texas (latitude $32°0.5'$ N and longitude $98°02'$ W). On aerial photographs, the structure is seen as a distinct photoanomaly approximately 3 km in diameter (Fig. 1). Arcuate tree lines and drainage patterns can be seen to surround a lighter-colored circular area 1 km in diameter. Stereoscopic views of the structure reveal that the concentric drainages occur within a series of ring-like troughs which surround a central uplifted area. A subtle circle 9 km in diameter surrounding the 3 km-diameter uplift and ring graben complex appears on Landsat images, but no corresponding structural disturbance has been recognized on the ground.

In the region, Cretaceous sedimentary strata are nearly horizontal, dipping less than 1° to the southeast. The upper 24 m of the Glen Rose Formation, a 60 meter-thick sequence of micritic and fossiliferous limestones alternating with less resistant marls, is exposed in river and stream beds in the area. The Paluxy Formation, a 15–20 meter thick, reddish-brown friable sandstone with local hematite concretions, crops out on valley slopes. The Walnut Formation, a 40 meter-thick sequence of calcareous clays and thin-bedded fossiliferous limestones overlies the Paluxy Formation. Lower and middle Walnut formation outcrops 17–20 m thick cap ridges between major drainages near Hico.

Distinctive lithologies in each formation serve as marker beds useful for structural mapping. A horizon of orange to brown, aphanitic, ferruginous limestone with manganese dendrites on a weathered surface occurs in the upper Glen Rose Formation less than 12 m below the contact with the Paluxy Formation. A flaggy unit of resistant, fine-grained calcite-cemented sandstone occurs in the lower Paluxy Formation, approximately 1 m above the Glen Rose Formation contact. A ripple-marked limestone bed (wavelengths about 0.5 m) overlying a white unconsolidated packsand occurs in the lower Walnut Formation at the Paluxy-Walnut formation contact. Less than 10 m above this sequence is a *Gryphaea* sp. zone which is excavated locally for aggregate. A white marl containing fragments of molds of *Oxytropidoceras* sp. overlain by a 1 meter-thick *Gryphaea* sp. zone marks the middle Walnut Formation.

It is likely that rock units which do not normally crop out in the Hico area, the lower beds of the Glen Rose Formation and perhaps silts, clays, and sands of the underlying Twin Mountains Formation, are uplifted near the center of the Hico Structure to lie immediately beneath a cover of Quaternary alluvium.

2. Surface Geology

Central Uplift

Plane table mapping at a scale of 1:2,400 (Wiberg, 1981) shows that the central area of the Hico Structure consists of an uplifted plug of folded sandstones of the Lower Paluxy Formation and limestones of the Glen Rose Formation. Fig. 2 shows that these folded strata probably occur in irregular fault-bounded blocks from 25 to 200 m across. Folds on the flanks of the uplift are undulating "pie crust-type" folds containing the youngest rock units in the uplift (unit "a" of Fig. 2). These gentle folds are best preserved about 200 m south of the houses in Fig. 2. Tighter folds occur nearer the center of the uplift. The axial traces of these tighter chevron-type folds, which show up well on the inset to Fig. 2 (lower left), are generally radial. Folds of this tighter type, containing vertical beds of upper Glen Rose limestone, have been mapped in detail (see especially the folds near

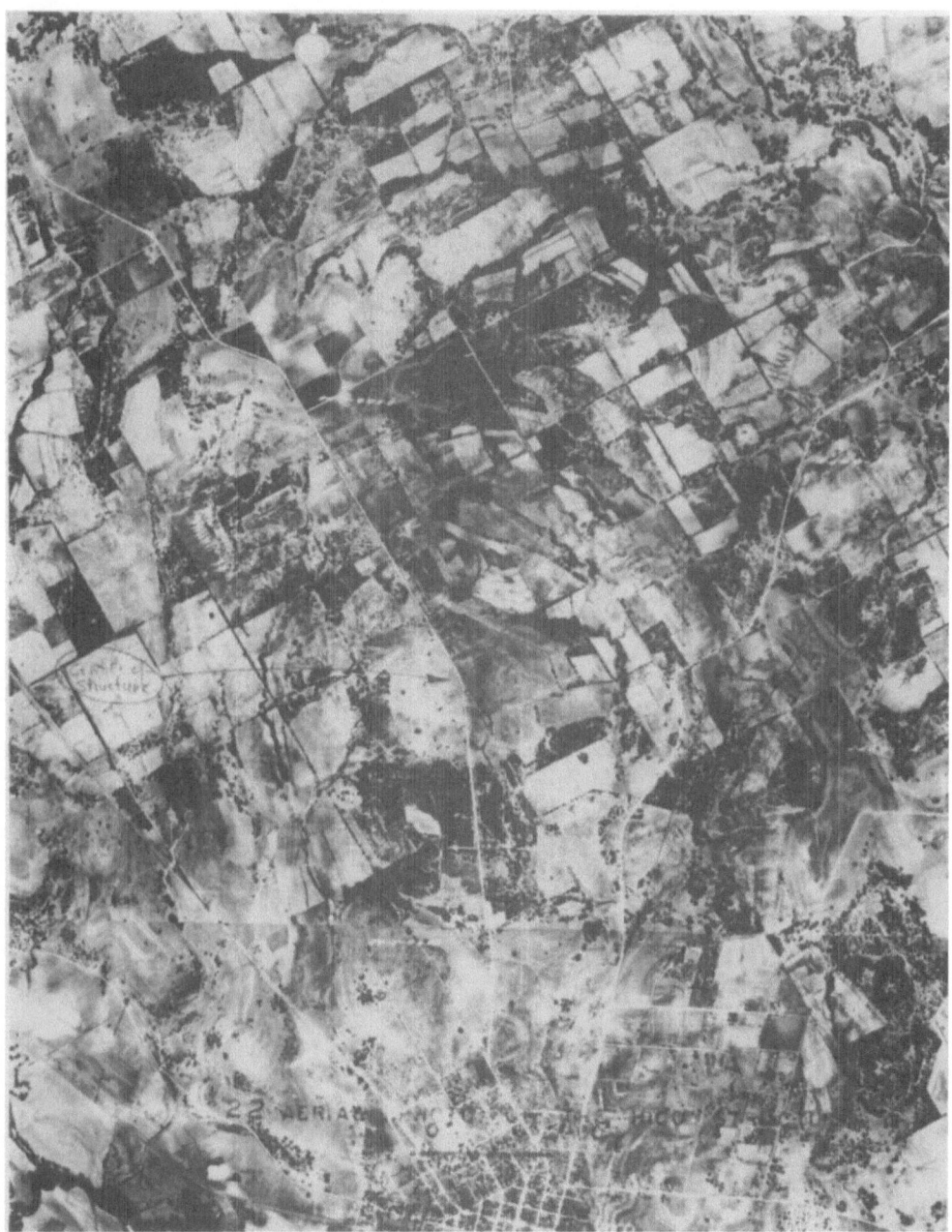

Fig. 1
Aerial photograph of the core of the Hico Structure. Annotations reproduced on photograph are by
W. J. McBride, early investigator of the structure. Tobin Aerial Surveys, San Antonio, Tx, No. 2137N,
Mosaic 16N-16E-2.

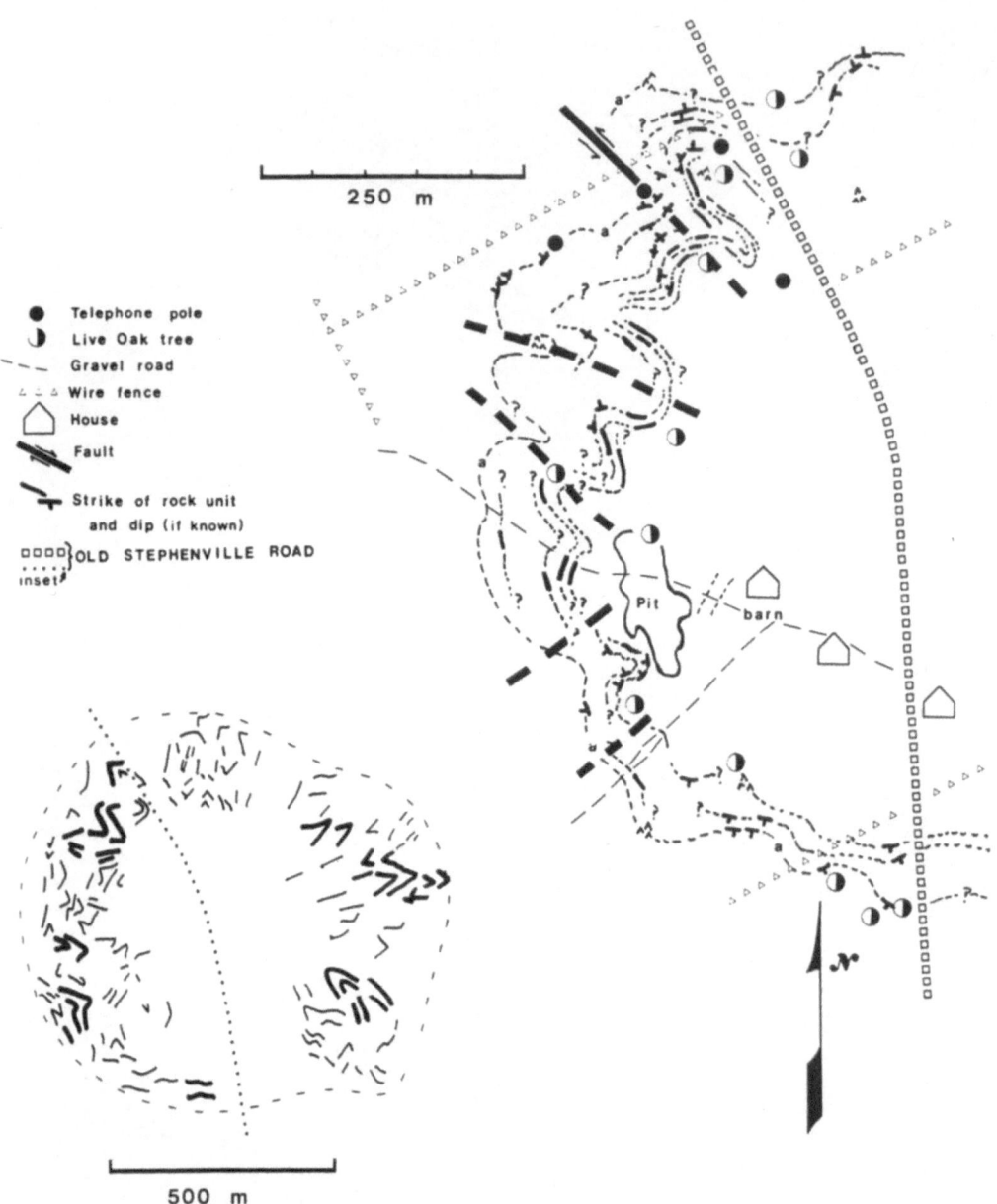

Fig. 2

The general pattern of folding at the central uplift of the Hico Structure (inset, lower left) and geologic map of the western half of the central uplift. Inset sketch map drawn from form lines on aerial photograph. Heavy lines mark clearly visible outcrops. Details of stratigraphy in Wiberg (1981). Unit "a" is lower Paluxy Formation sandstone; other rock units are Glen Rose Formation limestone.

the telephone poles in Fig. 2). Faults bounding the fold-containing central uplift blocks also appear to be sub-radial in the uplift.

In general, the age of uplifted rock layers increases consistently towards the center of the uplift. The rock layers within the innermost parts of the central uplift have not been positively identified. Nevertheless, estimates of total maximum uplift within the central uplift can be extrapolated from uplift values for the most deep-seated rock layer which has been positively identified within the upper Glen Rose Formation. This layer, a ferruginous limestone marker bed incorporated in chevron folds toward the outer flanks of the uplift, is at least 16 m above its regional plane. If limestones of the lower Glen Rose Formation occur inward from this layer, and sandstones of the Twin Mountains Formation are proven to occupy the extreme core of the central uplift (as shown in Fig. 4), then the total maximum uplift must be at least 80 meters. This total uplift value agrees closely with projected uplift values for the Hico Structure if the concentric pattern of vertical displacement within the central uplift of the Decaturville Structure (Offield and Pohn, 1979) is scaled to the Hico Structure. Details of this comparison, using a scaling ratio of 1:3 are given on pages 64—66 in Wiberg, 1981.

Ring Graben

A ring graben surrounds the megabrecciated central uplift. The inner graben boundary is somewhat obscure, probably a system of faults of small displacement. The outer boundary is a major ring fault (Fig. 3). Downdrop determinations in the ring graben are based on the relative level of marker horizons in the Walnut Formation. Mapping at the ring graben at a scale of 1:12,000 (Wiberg, 1981) shows that downdrop along the ring fault is 18 meters and 8 meters at locations 1.25 km south and 1 km east-northeast of the uplift center.

The Walnut Formation caps ridges that form a broken mid-graben annulus between concentric troughs floored by sandstones of the Paluxy Formation (Fig. 3). The structural relation of the annular troughs and ridge is uncertain. Fig. 4 shows the Walnut Formation outcrop as an outlier produced entirely by selective erosion adjacent to the ring fault (shown in Fig. 3) and to uplift flanks. Evidence of any faulting or folding within the troughs, however, is not readily apparent because the sandstone unit flooring them has been concealed by alluvium. Alternately, the Walnut-capped ridge may be an annular horst, or more precisely, a graben segment downdropped less than adjacent sections.

Walnut Formation limestones within southern and northwestern sections of the graben show plunging folds with circumferential axes (Fig. 3). These folds may be part of a circumferential fold belt. A series of irregularly shaped fault-bounded blocks within the graben have been downdropped to varying degrees. Differential downdrop of graben segments appears to have chopped the circumferential fold belt into even smaller irregular sections (Fig. 3). Reconnaissance surface mapping is not adequate to make more detailed statements about the circumferential folds or about the exact location or attitude of specific faults bounding individual graben segments.

3. Geophysics

Gravity and magnetic profiles (Wiberg, 1981) across the core of the Hico Structure reveal no excess or deficient subsurface mass or magnetic anomalies. Residual Bouguer anomalies on gravimetric profiles ranging from 0.5 mgal to 1.0 mgal correlate in several locations with the ring fault which bounds the graben and with a major north-south oriented fault within the eastern section of the ring graben.

Correlated electric logs of five oil and gas wells near the Hico Structure (one about 2 km north and one about 2 km west of the center of the structure) provide limited information on the subsurface geology. Casings are routinely set from the surface to near the Pennsylvanian-Cretaceous unconformity. Only one well records information shallow enough to locate the base of the Twin Mountains, the lowest Cretaceous formation. All that can be stated is that disturbance just beyond the ring graben must be shallower than 150—200 meters because the 1000 meter thick sequence of Pennsylvanian shales shown on the logs is undisturbed (Wiberg, 1981).

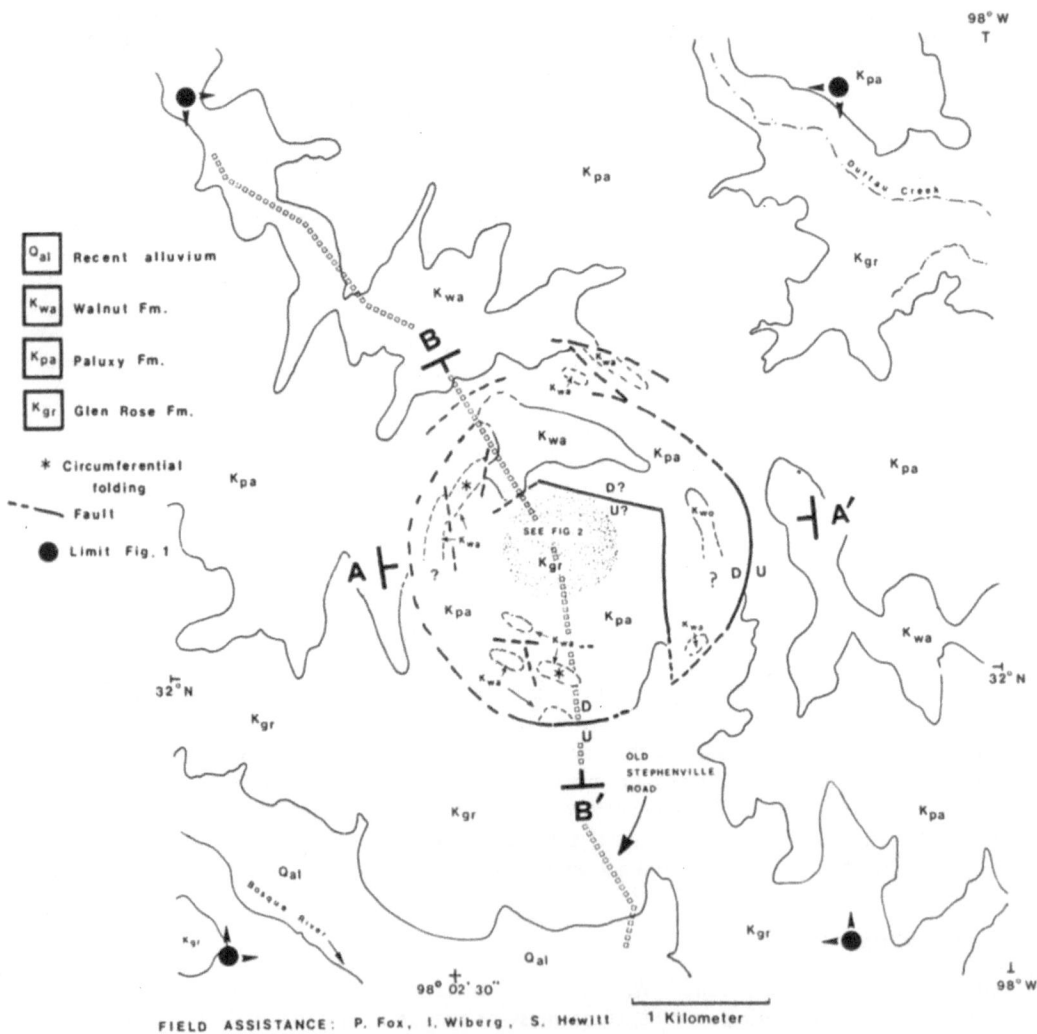

Fig. 3

Geologic map of the Hico Structure showing details of ring graben faulting and an annular mid-graben Walnut Formation outcrop pattern. Asterisks locate circumferential folds. Cross-Sections shown in Fig. 4.

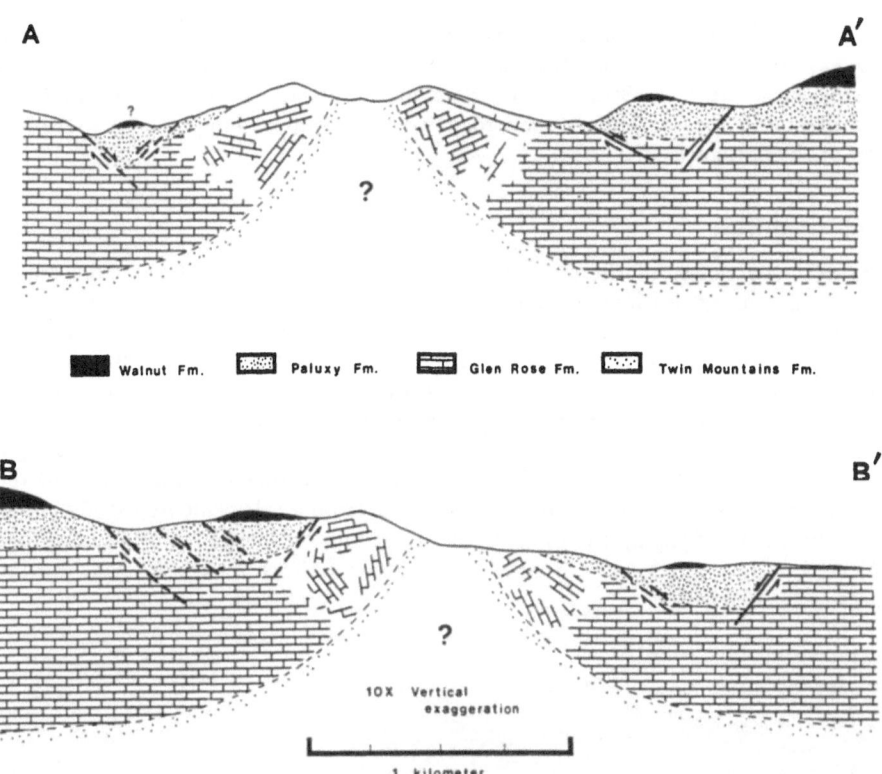

Fig. 4

East-west and north-south geologic cross-sections through Hico Structure. Uplift of Twin Mountains Formation and fault dip are conjectural.

4. Age and Erosional Modification

The Hico Structure is post-Cretaceous and pre-Pleistocene in age. The youngest deformed stratum in the structure is the Walnut Formation, of lower Cretaceous age.

Southwestern and northeastern parts of the Hico Structure have been severely eroded by the Bosque River and its tributary, Duffau Creek (Fig. 3 and Fig. 4). Nevertheless, the ring fault is still preserved and outliers of downdropped Walnut Formation still remain in the south. Erosion has also reduced the core of the central uplift to a topographic depression (Fig. 4, A-A') and alluvium conceals the unidentified bedrock units. A south-flowing drainage system originating in the central depression of the uplift (perhaps oriented along a prominent radial fault) has eroded away much of the southern section of the central uplift (Fig. 4, B-B'), modifying the once ring-like inlier into a horse-shoe opening southward.

The Hico Structure apparently pre-dates the development of cross-cutting joints that controlled the establishment of the course of the Bosque River system. The joints, seen as lineaments 3 to 10 km long on Landsat images, are fractures which propagated upward from subsurface Precambrian and Paleozoic features at an unknown time (p. 24–25, Wiberg, 1981). Lineaments cutting across the northeast periphery of the Hico Structure (which have strongly influenced the drainage course of Duffau Creek 6 and 12 km north and 6 km east of the center of the structure) are as straight and undisturbed as lineaments far removed from the Hico Structure. Landsat images also show that the Bosque River follows a distinctly arcuate course concordant with the outer 9 km-diameter southwest perimeter of the structure.

5. Shock Features

Exposure near the center of the structure, where shock features might be expected, is poor except in two excavations. In one of these (about 200 meters east of the eastern-most house shown on Fig. 2) the friable limestone shows surfaces up to 40 cm long with clearly divergent striations (Fig. 5). Although the surfaces along which the striations are developed are irregular, rather than clearly conical as in shatter cones in harder rocks at other structures, the surfaces are confidently interpreted as shatter fractures.

Other evidence of impact such as small-scale brecciation, planar shock features and high-pressure silica polymorphs, were not found. The quartz-bearing material closest to the center of the structure available for examination was from a calc-cemented sandstone of the Paluxy Formation fringing the uplift (unit "a" of Fig. 2). Quartz from this sandstone was subjected to acid treatment and x-ray diffraction in a search for coesite or stishovite, but results were negative.

Fig. 5
Shatter cone on borrow pit floor in uplifted Glen Rose Formation limestone at central uplift of Hico Structure.

6. Analogous Impact Structures

Most ancient and eroded impact sites large enough to be preserved are craters of the complex type (Grieve, 1982), i. e. impact features with a prominent central uplift, or if large enough, with a concentric multi-ring configuration. In general, a central uplift and surrounding depressed zone (Table I) as present at the Hico Structure, is found at eroded impact features in sedimentary sequences varying in diameter from 4 to 25 km such as the Sierra Madera Structure (Wilshire *et al.*, 1972), Decaturville Structure (Offield and Pohn, 1979), Flynn Greek Structure (Roddy, 1977), Kentland Structure (Laney and

Table I: A structural comparison of the Hico Impact Structure to other impact features in sedimentary strata.

		Hico	Sierra Madera	Decaturville	Flynn Creek	Kentland	Wells Creek	Gosses Bluff
diameter of complete structure (km)		3 (+?)	13	9	3.8	12.4	13	25
central uplift	diameter (km)	< 1	8	2.4	< 1	4[b]	> 5	3.5
	amount of uplift (m)	80 (+?)	> 1200	300−500	450	600	< 800	3000
	folding present	major	major	moderate	moderate	minor	minor	very minor
ring graben	width (km)	1	0.8−1.6	1.6	1	< 2	4	10
	circumferential folds present	yes	yes	yes	yes[a]	no	no	no
	concentric fracture system	yes	yes	yes	yes	no	yes	complex

(a) ring anticline on crater floor seen in drill holes
(b) intensely deformed 1 km-diameter core within 4 km-diameter central uplift

Van Schmus, 1978), Wells Creek Structure (Wilson and Stearns, 1968) and Gosses Bluff Structure (Milton *et al.*, 1972). Strata in the central uplifts of these impact structures have been uplifted from 300 meters to 3000 meters compared to the 80 meters estimated at the Hico Structure (Table I).

The degree and type of folding and faulting seen at the Hico Structure is grossly compatible with deformation seen at the other impact structures (Table I). Some degree of folding is present within the central uplifts of all the analogous impact structures in Table I, but the Hico Structure appears to be at the "predominantly intensely folded" end of the continuum together with the Sierra Madera Structure, and to a lesser degree the Decaturville and Flynn Creek Structures. Also, the Hico, Flynn Creek and Sierra Madera Structures have vertical beds and radial fold axes within central uplifts. Finally, both the Hico and Sierra Madera Structures have gentler, more sinuous and undulating concentric folds on central uplift flanks.

Circumferential folds within the ring graben are better developed at the Flynn Creek, Decaturville and Sierra Madera impact structures than at the Hico Structure. Single ring faults or a series of concentric faults surround the uplifts Decaturville, Wells Creek and Sierra Madera impact structures. The ring graben at the Hico Structure may have structural counterparts at the Decaturville and Wells Creek structures where an inner graben, horst and outer graben also surround central uplifts.

The close similarity of the style of structural deformation and the shatter fracturing seen at the Hico Structure to that at accepted impact structures where more complete geological information is available indicates that the Hico Structure should be added to the list of terrestrial impact structures.

7. Suggestions for Further Study

Further detailed mapping within the central uplift will determine whether the folding preceded the radial faulting of the uplift into irregular blocks. Coring of the depression at the center of the uplift should allow an exact maximum uplift value to be calculated from a positive identification of the oldest and originally most deep-seated stratigraphic units within the core of the uplift.

Efforts to better determine the age of the structure should involve attempts to better date the cross-cutting joints, together with more detailed field mapping to look for younger unconformable sediments and attempts to reconstruct the original crater morphology, using local erosion rates to give a minimum age.

A closer comparison with other impact structures may show that the variability of central uplift folding reflects the rheology of the target rock under different impact circumstances. The predominance of folding within certain central uplifts may give a clue to the relative depth of a specific rock layer beneath the initial energy surface — intensely folded layers being closer to the original impact surface. On the other hand, uplifts like at Hico, which are characterized by ductile deformations, may have formed in semi-lithified sediments (although the formation of a shatter cone at the Hico Structure indicates some degree of consolidation).

Acknowledgements

Oscar Monnig, an amateur astronomer of Fort Worth, Texas, called attention to the Hico Structure as a possible impact structure in the late 1960's. Mary McBride provided unpublished reports prepared by her late husband William J. McBride for Humble Oil and Refining Company while investigating the Hico Structure as an oil prospect in the 1950's. Baylor University kindly provided aerial photographs. I am especially grateful to the family of Dr. Richard Lysiak, physics professor, TCU, for letting me use their ranch house as a field base. Others who provided helpful on-site impressions and comments include: J.R. Underwood, Jr., A.J. Ehlmann, R. Wetterauer, F. Hörz, D. Amsbury, R. Dietz, P. Lambert, C.R. Seeger and C.L.V. Aiken. The manuscript was reviewed by Odette B. James and D.J. Roddy, whose long-distance encouragement is much appreciated. My husband, Dan Milton, found the shatter cone. Without his persistent and good natured support during a particularly busy time for both of us, the completion of this manuscript would not have been assured.

References

Grieve, R.A.F., 1982: The record of impact on Earth: implications for a major Cretaceous/Tertiary impact event. In: Geological Implications of Impacts of Large Asteroids and Comets on the Earth (Silver, L.T. and P.H. Schultz, eds.). Geol. Soc. Amer. Spec. Paper 190, 25–37.

Laney, R.T., and W.R. Van Schmus, 1978: A structural study of the Kentland, Indiana, impact site. Proc. Lunar Sci. Conf. 9th, Pergamon Press, New York, 2609–2632.

Milton, D.J., B.C. Barlow, Robin Brett, A.R. Brown, A.Y. Glikson, E.A. Manwaring, F.J. Moss, E.C.E. Sedmik, J. Van Son, and G.A. Young, 1972: Gosses Bluff Impact Structure, Australia. Science, 175, 1199–1207.

Offield, T.W, and H.A. Pohn, 1979: Geology of the Decaturville impact structure, Missouri. U.S. Geol. Surv. Prof. Paper 1042, 48 p.

Roddy, D.J., 1977: Pre-impact conditions and cratering processes at the Flynn Creek crater, Tennessee. In: Impact and Explosion Cratering (Roddy, D.J., R.O. Pepin and R.B. Merrill, eds.). Pergamon Press, New York, 277–308.

Wiberg, L., 1981: The Hico Structure: a possible impact structure in north-central Texas, U.S.A. M.S. Thesis in geology, Texas Christian University, Fort Worth, Texas.

Wiberg, L., 1980: The Hico Structure: a possible astrobleme in north-central Texas, U.S.A. In: Papers presented to the conference on multi-ring basins: formation and evolution. Lunar and Planetary Institute, Houston, 112–114.

Wilshire, H.G., T.W. Offield, K.A. Howard, and D. Cummings, 1971: Geology of the Sierra Madera cryptoexplosion structure, Pecos County, Texas. U.S. Geol. Surv. Paper 599-H, 42 p.

Wilson, C.W., and R.G. Stearns, 1968: Geology of the Wells Creek Structure, Tennessee. Division of Geology, Bull. 68, 236 p.

Book Review

Silver, L. T. and P. H. Schultz
Geological Implications of Impacts of Large Asteroids and Comets on the Earth.
Geol. Soc. Amer. Spec. Paper 190, 528 pp.

This comprehensive volume contains the proceedings of a conference on the same topic held at Shnowbird, Utah in 1981. The purpose of the conference was to bring together scientists from all disciplines which could contribute to the problem of the relationship between large extraterrestrial body impacts and terrestrial evolution, especially biological mass extinctions. The conference was attended by a spectrum of scientists ranging from astronomers, geochemists, impact specialists, climatologists, and biologists to pale-ontologists. A similar scientific bandwidth is covered by the 48 contributions in this amazing volume.

In the first part of the book, more general topics are addressed by the authors. The first articles deal with the record and rate of impact cratering on Earth, including estimates from the astronomically observable bodies which could collide with the Earth, and from the more than 100 known impact structures on Earth. The following series of contributions describe experiments and theoretical considerations on the impact process of large bodies in continental and oceanic areas. Possible atmospheric and climatological short- and longterm effects of large impacts, which are considered to be a major cause of biological mass extinctions, are discussed. The problem of projectile identification by the analysis of characteristic trace elements such as Ir is addressed in several papers. Several contributions in this first part of the book deal with the general problem of biological mass extinctions: How good are they documented? What are the errors and gaps in sampling the fossil record? How catastrophic were extinctions? etc.

In the second part of the book, the evidence for an important impact event in numerous marine and terrestrial sites distributed worldwide and the evidence for and against a catastrophic mass extinction at the Cretaceous/Tertiary boundary is presented and discussed in detail. In the final articles some evidence for other phanerozoic extinctions and impact signatures is given.

The book is an extremely useful and well presented collection of articles on most aspects of the Cretaceous/Tertiary boundary problem and the impact hypothesis. It is an excellent introduction to the problem and a good basis for understanding the importance and bearing of the proliferating articles on this subject.

Jean Pohl, Munich

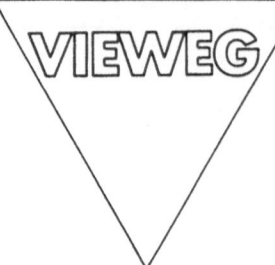

Rodney A. Gayer (Ed.)

The Tectonic Evolution of the Caledonide-Appalachian Orogen

1985. VI, 194 pages, 45 fig. 16 x 24 cm. (Earth Evolution Sciences. International Monograph Series on Interdisciplinary Earth Science Research and Applications, ed. by Andreas Vogel.) Softcover

The geology of the Caledonide-Appalachian orogen is probably the most intensively studied of all mountain chains, and yet its origins and evolution are still highly controversial. Interest in the orogen has been heightened in recent years by a vast amount of new information arising from work in connection with I. G. C. P. project no. 27 – the Caledonide Orogen. It is clear that, in addition to modelling the Caledonide-Appalachian orogen on present plate boundaries, it is necessary to recognise the importance of major fault contacts. As with the Western Cordillera of North America, where stratigraphic, structural and palaeomagnetic studies have demonstrated the presence of a large number of suspect terranes, juxtaposed by major strike-slip displacements, so the Caledonide-Appalachian orogen is best investigated by terrane analysis. Each terrane is regarded as fault displaced unless a definite connection between adjacent terranes can be established.

The articles presented in this thematic issue have been carefully selected to give a wide coverage of the orogen, reviewing the important facets of its evolution in terms of plate tectonics and suspect terrane analysis. Such an approach, of necessity, covers a wide spectrum of earth science disciplines but every effort has been made to integrate the individual articles into a general framework. Thus the first article on British Caledonian terranes lays the foundations for the main reviews and the final article presents a coordinated tectonic model for the evolution of the belt, integrating the material from the individual contributions. The account presented in this issue represent an up to date overview of the evolution of one of the most intriguing of orogenic belts.